Special Topics in Dam Engineering

Edited by Hasan Tosun

Published in London, United Kingdom

Special Topics in Dam Engineering
http://dx.doi.org/10.5772/intechopen.97978
Edited by Hasan Tosun

Contributors
Hasan Tosun, Gabriella Vaschetti, Jonathan A. French, Angélica Luciana Barros de Campos, Ruben Jose Ramos Cardia, Welitom Ttatom Pereira da Silva, Marla T.B. Geller, Anderson Alvarenga de Moura Meneses, Tamaz Chelidze, Teimuraz Matcharashvili, Ekaterine Mepharidze, Levan Mebonia, Mirian Kalabegashvili, Nadezhda Dovgal

First published in London, United Kingdom, 2022 by IntechOpen
IntechOpen is the global imprint of INTECHOPEN LIMITED, registered in England and Wales, registration number: 11086078, 5 Princes Gate Court, London, SW7 2QJ, United Kingdom

British Library Cataloguing-in-Publication Data
A catalogue record for this book is available from the British Library

Additional hard and PDF copies can be obtained from orders@intechopen.com

Special Topics in Dam Engineering
Edited by Hasan Tosun
p. cm.
Print ISBN 978-1-80355-979-7
Online ISBN 978-1-80355-978-0
eBook (PDF) ISBN 978-1-80355-980-3

For EU product safety concerns:
IN TECH d.o.o., Prolaz Marije Krucifikse Kozulić 3, 51000 Rijeka, Croatia,
info@intechopen.com or visit our website at intechopen.com.

Meet the editor

Hasan Tosun is the president of Turkish Society on Dam Safety and a former researcher (full professor) in the Civil Engineering Department, Eskisehir Osmangazi University, Turkey. As director and dean, he has governed the Earthquake Research Centre for twelve years and the Agricultural and Life Faculty for three years, respectively. He was the vice rector at Uşak University, Turkey, and dean of the Engineering Faculty there. He specializes in geotechnical issues for earth and rockfill dams. Up to 1997, he worked at the General Directorate of State Hydraulic Works and supervised the geotechnical studies of large dams constructed in Turkey. He has published more than 320 technical papers published in national and international journals and conference proceedings. Currently, he is the rector of Mudanya University, Turkey.

Contents

Preface

Water is one of the most basic resources for living a healthy life. It is integral to human health as well as agriculture, energy production, and more. Energy can be produced from other sources; however, water has no alternative in terms of providing domestic and industrial energy needs and for irrigation purposes. For this reason, cities and countries are trying to use these limited natural resources correctly and ensure the principle of continuity in use. This means that storing water for later use is an important aspect of dam engineering. As such, the number of reservoirs and related structures (dams and irrigation systems) constructed for the purpose of storing fresh water has significantly increased worldwide. Many large projects related to the effective use of freshwater have been implemented, especially in developing countries.

Land resources in which agricultural activities can be carried out in the world are also restricted. Food production, which is one of the most basic needs of social life, is only possible with the continuous execution of agricultural activities. There are about 27 million square kilometers of arable land in the world. Countries carry out irrigation projects to increase the yield of these agricultural areas and try to meet their food needs using advanced technology. Unfortunately, natural and artificial hazards increase food demands.

Recent studies indicate that water and earth resources are unevenly distributed throughout the world. There is also an imbalance in usage rates of freshwater worldwide. In developing countries, agricultural water use is about 75 percent of total global consumption, while industrial use accounts for about 20 percent. The remaining 5 percent is used for domestic purposes. Additionally, two of every three people living on Earth will experience water stress by 2025, according to recent projections. Scientists state that clean water supplies and sanitation are likely to become major problems in many parts of the world. Water-borne diseases resulting from water pollution continue to be a major cause of illness in undeveloped and developing countries.

This book consists of six excellent contributions to dam engineering. The chapters cover a wide range of topics about dam issues. Chapter 1 presents a study that was performed on dams within a cascade system posing a high total risk to downstream life, such as the large dams on the Euphrates River in Turkey. Chapter 2 discusses the use of a geomembrane system in dam engineering and presents projects in this area from around the world. Chapter 3 deals with managing impounded water quality and serves as a literature survey of management and environmental issues with case studies. Chapter 4 presents the results of an investigation performed on the possibility of a cascade failure, based on Brazilian dams. Chapter 5 focuses on uncertainty factors for the lifecycle assessment of hydroelectric power plants. Finally, Chapter 6 discusses the potential of nonlinear dynamics tools in the real-time monitoring of large dams using the case of the Enguri high-arch dam.

As an editor, I am thankful to the chapter authors for their contributions. I hope this book proves to be a useful resource for readers studying dam engineering.

Dr. Hasan Tosun
Professor,
Mudanya University,
Bursa, Turkey

Chapter 1

Recent Evaluation on Total Risk of Cascade Dams on Murat River of Upper Euphrates Basin, Turkey

Hasan Tosun

Abstract

The dams within a cascade system pose a high total risk to the downstream life, even if they provide significant benefits in terms of flood protection, irrigation water, and domestic water supply and energy production; a dam in a cascade system also poses a substantial risk from the point of view of other structures in the basin and causes the danger to grow due to the triggering effect from the point of view of dam failure. In this study, the total hazard of the dams in the Murat River located in the upper part of the Euphrates-Tigris Basin, the largest basin in Turkey, will be evaluated, and calculations made about it will be summarized. The possible hazards in a cascade system will be highlighted. Ten large dams of various types ranging from 36 m to 138 m in height from the river basin have been considered in this context. The analysis results show that six dams are under near-source effect in terms of seismicity, and all of the dams considered have a high total risk, although they have different hazard ratios. In addition, three separate dams located within the cascade structure carry a much greater risk regarding the dangers that other structures may create.

Keywords: cascade system of dam, dam failure, overtopping, total risk

1. Introduction

There are two fundamental factors contributing to the total risk of the dams: (1) the structure's risk rating and (2) the rating of the dam site on seismic hazard. In general, the peak ground acceleration (PGA) is utilized to represent the design earthquake for a dam site. The risk rating is defined on the basis of the reservoir capacity, dam height, potential damages to the downstream, and evacuation requirements. While the International Commission on Large Dams (ICOLD) considers the seismic and risk ratings separately, the Bureau's method combines these factors to compute the total risk factor (TRF) for dams [1, 2]. Later, ICOLD introduced the guidelines for selecting seismic parameters for large dams [3].

Turkey has 26 water basins. The largest one in the country is the Euphrates-Tigris basin, which has two main rivers and about 53 km^3 per year of water yield. The first is the Tigris River, which crosses the Iraq border and joins the Euphrates River. Several large dams exceed 10 in the Tigris River basin for utilizing earth and water resources. The second one, namely the Euphrates River, located west of the basin, is 2800 km

#	Dam	River	Height from river bed (m)	Function*	Type**	Owner***	Completed year****	Body volume (hm^3)	Reservoir capacity (hm^3)	Installed capacity (MW)	Irrigation area (ha)
1	Agri-Yazici	Altincayir	84	I + WS	EF	S	2001	7.92	202.9	—	25,079
2	Alparslan-1	Murat	88	E	RF	S	2009	3.78	2993.0	160	—
3	Alparslan-2	Murat	108	E + I + WS	AEF	P	2020	12.45	2431.0	280	78,000
4	A.Kalekoy	Murat	88	E	C	P	2020	5.27	516.5	500	—
5	Beyhan-I	Murat	84	E	RCC	P	2015	1.22	369.1	582	—
6	Beyhan-II	Murat	42	E	CG	P	U/C	0.23	78.9	264	—
7	Gayt	Gayt	36	I	EF	S	1998	0.44	23.0	—	4200
8	Gulbahar	Kocan	66	I	EF	S	2003	1.48	14.1	—	1572
9	Patnos	Gevi	38	I	EF	S	1992	1.30	33.4	—	5973
10	Y.Kalekoy	Murat	127	E	RCC	P	2018	2.34	783.8	636.6	—

*E: Energy; I: Irrigation; and FC: Flood control.
**RF: Rockfill; EF: Earthfill; CG: Concrete gravity; and HF: Hardfill AEF.
***S: State; and P: Private.
****U/C=Under construction.

Table 1.
Physical properties of 10 dams under study [13].

long and joins the Tigris in Iraq, flowing into the sea (Persian Gulf). On the Euphrates River and its tributaries, 32 large dams have been planned to be built and used for energy and irrigation purposes. Several dams are completed on the main part of the Euphrates River, including Karkamis, Birecik, Ataturk, Karakaya, and Keban. There exist many studies related to these dams, which are reported in [4–12].

The Euphrates River has three main tributaries in the upper region of the basin: Karacasu, Perisuyu, and Murat rivers. The Keban dam lake is the confluence of these rivers. The Karacasu River is located further west and contains many large dams along with its other tributaries. The other main tributary, which is called as Perisuyu River, includes eight large dams called Kigi, Kalecik, Konaktepe, Ozluce, Pempelik, Seyrantepe, Tatar, and Uzuncayir. The third one is the Murat River, which originates in the northeast of Lake Van and flows westward, forming deep valleys to merge with the Karacasu tributary. There are 10 large dams which have been completed on this river and its tributaries including Ağrı-Yazıcı, Alparslan 1, Alparslan 2, Lower Kalekoy, Beyhan-I, Beyhan-II, Gayt, Gülbahar, Patnos, and Upper Kalekoy [13].

The 722-km Murat River originates from the north of Lake Van, flows toward the southwest, and reaches the Malazgirt Plain. This river receives other small tributaries in the Agri region on its way. After taking the Hinissu coming down from the Bingol Mountains, it enters the Mus Plain from the north by drawing hard geomorphologic features and taking the Karasu coming from Mount Nemrut here, then flows west, passing through narrow gorges and in front of Palu, merges with the Munzur-Peri water coming from the North, and immediately reaches the reservoir of Keban dam. The total hydraulic capacity is about 2650 MW. The dams and power plants with an installed capacity of 2149 MW have been planned on this river, and today more than 80% of them have been completed or are under construction. It is planned to irrigate about 120,000 ha of agricultural land in the region with the river's water.

In this paper, the total risks of 10 above-mentioned dams on the Murat River will be examined, and their location within the cascade system will be evaluated. In particular, the effects of structures located in a cascade system will be focused on in terms of total risk related to upstream and downstream facilities and potential hazard situations expressed separately for each dam. The main physical characteristics of these dams, whose heights range from 36 m to 138 m from the river bed, are presented in **Table 1**. These dams were built to generate energy, to supply drinking water, and for irrigation usage. The dams located on the main river and built for electrical energy generation are privately owned. The related dams have been constructed in different types. Conventional types of dams (rock- and earth-fill) belong to the state. One-half of these dams were built by the state and the other half by the private sector. State dams are usually for irrigation purposes, while private dams are entirely for electrical energy generation (**Table 1**).

2. Methodology and materials

For the seismic analyses, seismic sources, which are acting on the investigation area, were identified and evaluated on the basis of guidelines and the hazard modeling developed for the region to form seismic model of the dam sites [14, 15]. The study of seismic activity includes both the Deterministic Seismic Hazard Analysis (DSHA) and the Probabilistic Seismic Hazard Analysis (PSHA) [16, 17]. Some attenuation relationships, which were developed on the basis of actual earthquake data, were adopted to obtain the peak ground acceleration (PGA) on each dam site, because of being no enough strong motion records. The analyses performed using the DAMHA software, which is a specified program, were developed in Osmangazi University. This program was used in several studies [18–32].

The recurrence interval of earthquakes was estimated as based on the seismic sources identified previously. The earthquakes that occurred within the last 100 years were used for estimating seismic parameters. Throughout the study, seismic zones and earthquakes within the area having a radius of 100 km around the dam site were considered. For all analyses throughout this study, the peak ground acceleration was obtained by considering FEMA's earthquake definition by two methods [33].

Throughout this study, two methods have been considered for estimating the total risk for dam structures. In DSI guidelines, the total risk factor depends on reservoir capacity, height, evacuation requirement, and potential hazard [34]. The Bureau method, which considers the dam type, age, size, downstream damage potential, and evacuation requirements, was utilized to realize the risk analyses of the basin. It suggests four separate risk classes ranging from I (low risk) to IV (extreme risk) based on the total risk factor (TRF).

3. Analyses

The area under investigation has a very complex geology and very active seismicity. The author of this paper and their colleagues introduce more comprehensive information about the geology and seismotectonic structure of the region [11, 12]. Two main energy sources are acting in the region: The North Anatolian Fault Zone (NAFZ) and The East Anatolian Fault Zone (DAFZ). The NAFZ is one of the best-known strike-slip faults in the world because of its significant seismic activity and well-developed surface features. It is approximately 1500 km long and extends from eastern Turkey in the east to Greece in the west. The width of NAFZ ranges from a single zone of a few hundred meters to multiple shear zones of 40 km. The DAFZ has a 550-km length and is a northeast-trending, sinistral strike-slip fault zone. It comprises a series of faults arranged parallel to the general trend. It is a transform fault forming parts of boundaries between the Anatolian and Arabian Plates [5].

In this region, there are many earthquakes with a magnitude greater than 4.0 (based on the surface wave magnitude scale (Ms)), which indicates that the region is very active. Furthermore, there were only two earthquakes with a magnitude equal to or greater than 7.0 in the basin. The first one is the 1939 Erzincan earthquake with M_s of 7.9 and the second is the 1949 Karliova earthquake with M_s of 7.0. Earthquakes occurred in the region, and prevailing seismotectonic structures are given in **Figure 1**.

For seismic hazard analyses of 10 dam sites, a detailed study was performed to identify all possible seismic sources based on the macro seismotectonic model of Turkey. The National Disaster Organization and other institutes prepared the map for general use. However, later it was modified by the author and his colleagues to be used for dam projects. The region's seismic history and local geological features were used to quantify the rate of seismic activity in the basins. The seismic stability of dams considered in this study was re-analyzed according to the new map on seismic

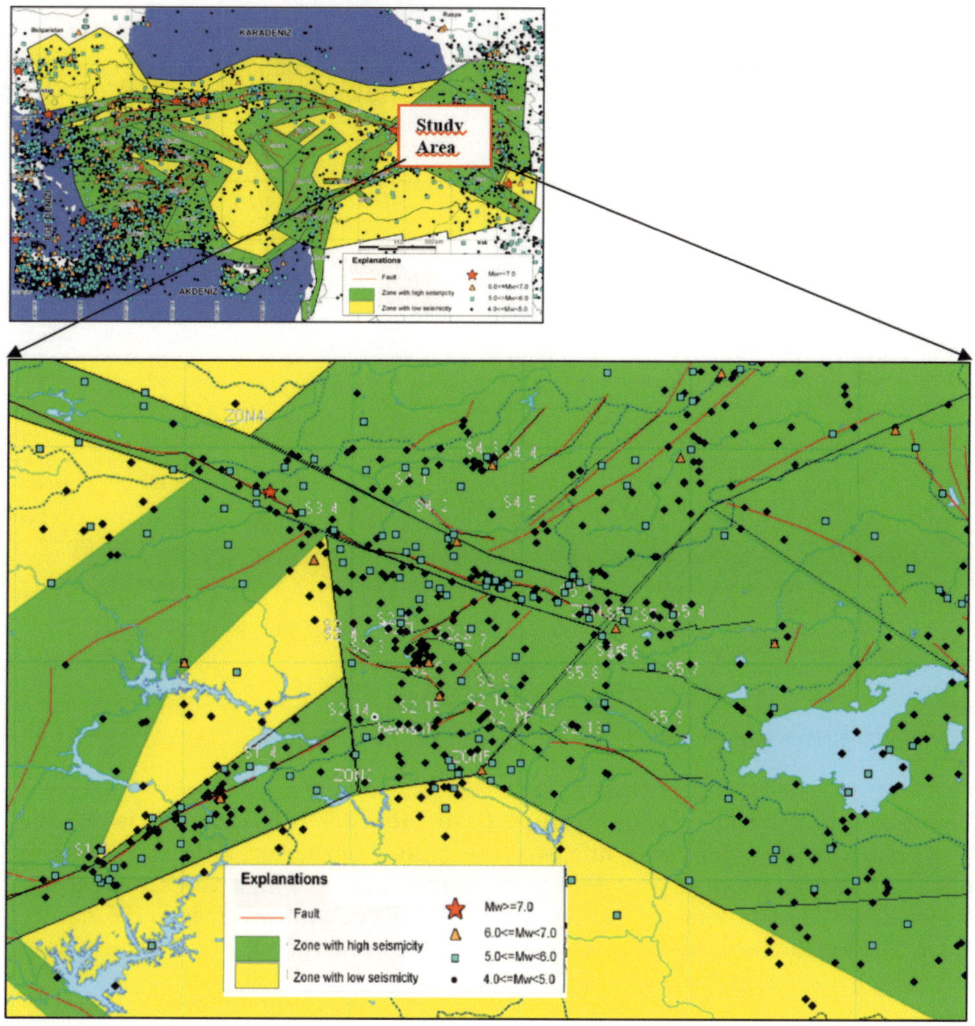

Figure 1.
Earthquakes on the seismotectonic model of the region.

zonation and updated national regulations on seismic design. The results of seismic hazard analyses are given in **Table 2**.

The seismic hazard analyses performed using two different methods indicate that the PGA value varies within a wide range for the dam sites. The Deterministic Seismic Hazard Analysis (DSHA) shows that the PGA values for the 50 percentile range from 0.107 to 0.366 g, while those for the 84 percentile are between 0.174 and 0.596 g. According to Probabilistic Seismic Hazard Analysis (PSHA), peak ground acceleration (PGA) changes within a wide range for all earthquakes levels. These values mainly depend on the position of dam sites and the nature of predictive relationships. Beyhan-II dam has a PGA value of 0.872 g, which is the maximum value for all dams considered in this study based on the Safety Evaluation Earthquake (SEE). Beyhan-I and Yukarı Kalekoy dam sites will also be subjected to relatively high PGA values because of their proximity to the energy source. The sites of Gayt, Gülbahar, and Alparslan1 dams also pose high PGA values.

#	Dam	Deterministic Method*				Probabilistic Method**		
		M_{max}	R_{min} (km)	Mean PGA + 50% in g	Mean PGA + 84% in g	OBE in g	MDE in g	SEE in g
1	Agri-Yazici	6.6	29.1	0.108	0.175	0.226	0.280	0.352
2	Alparslan-1	6.4	8.9	0.261	0.422	0.378	0.482	0.633
3	Alparslan-2	6.4	13.2	0.197	0.319	0.344	0.429	0.552
4	A.Kalekoy	6.4	12.7	0.200	0.323	0.287	0.363	0.474
5	Beyhan-I	6.3	3.0	0.338	0.551	0.466	0.604	0.808
6	Beyhan-II	6.3	1.7	0.366	0.596	0.507	0.655	0.872
7	Gayt	6.5	7.2	0.305	0.491	0.402	0.520	0.698
8	Gulbahar	6.4	9.2	0.251	0.406	0.355	0.454	0.601
9	Patnos	6.3	25.8	0.107	0.174	0.206	0.255	0.323
10	Y.Kaleköy	6.1	4.3	0.285	0.465	0.430	0.552	0.731

**M_{max} = Maximum earthquake magnitude in M_w; R_{min} = Minimum distance to fault segment; Mean PGA + 50% = Mean Peak Ground Acceleration at the 50th percentile; Mean PGA + 84% = Mean Peak Ground Acceleration at the 84th percentile.*

***OBE = Operational Based Earthquake; MDE = Maximum Design Earthquake; SEE = Safety Evaluation Earthquake.*

Table 2.
The results of seismic hazard analyses.

The results of total risk analyses are given in **Table 3**. The hazard ratio is very high for six dams and high for two, while two pose a moderate hazard ratio. According to ICOLD classification, all of them are categorized as a very high-risk ratio, except Patnos dam. This study shows total risk factor (TRF) values are between 136.2 and 204.4. The Bureau method indicates that all dams considered for this study are classified as high-risk ratios.

4. Discussion

Within the scope of this study, the total risks of 10 dams constructed on the Murat River and its tributaries and their impact on the cascade system were examined. There are six large dams within the cascade system of the main river. These structures are Alparslan-1, Alparslan-2, Upper Kalekoy, Lower Kalekoy, Beyhan-I, and Beyhan-II dams, from the upstream and the downstream, respectively. The positions and energy levels of these dams within the cascading structure are presented in **Figure 2**. In addition to these dams, there are four separate large dams (Agri-Yazici, Gayt, Gulbahar, and Patnos) located on the tributaries of Murat River, which are embankment dams and have relatively low storage.

The analyses indicate that Alparslan-1, Beyhan-I, Beyhan-II, and Upper Kalekoy dams, located in the cascading system, are under near fault motions. The relevant dams are 8.9, 3.0, 1.7, and 4.3 km away from the energy source, which can, on average, produce an earthquake of 6.4-moment magnitude (Mw) scale, respectively. Therefore, large PGA values appear in these dam sites, resulting in a hazard class with a "very high" ratio (**Tables 2** and **3**). These dams in the cascading system, with high total risk factors, are

#	Dam	PGA in g	M_{max}	Hazard Analysis		Total Risk (ICOLD,1989)			Total Risk (Bureau, 2003)		
				Class	Hazard Ratio	Risk factor	Risk class	Risk ratio	Risk factor	Risk class	Risk ratio
1	Agri-Yazici	0.175	6.6	II	Moderate	36	IV	Very high	136.2	III	High
2	Alparslan-1	0.422	6.4	IV	Very high	36	IV	Very high	148.2	III	High
3	Alparslan-2	0.319	6.4	III	High	36	IV	Very high	201.0	III	High
4	A.Kalekoy	0.323	6.4	III	High	36	IV	Very high	143.3	III	High
5	Beyhan-I	0.551	6.3	IV	Very high	36	IV	Very high	173.5	III	High
6	Beyhan-II	0.596	6.3	IV	Very high	36	IV	Very high	173.5	III	High
7	Gayt	0.491	6.5	IV	Very high	32	IV	Very high	204.4	III	High
8	Gulbahar	0.406	6.4	IV	Very high	34	IV	Very high	193.9	III	High
9	Patnos	0.174	6.3	II	Moderate	28	III	High	163.4	III	High
10	Y.Kalekoy	0.465	6.1	IV	Very high	36	IV	Very high	168.7	III	High

Table 3.
Total risk analyses of the dams considered for the study.

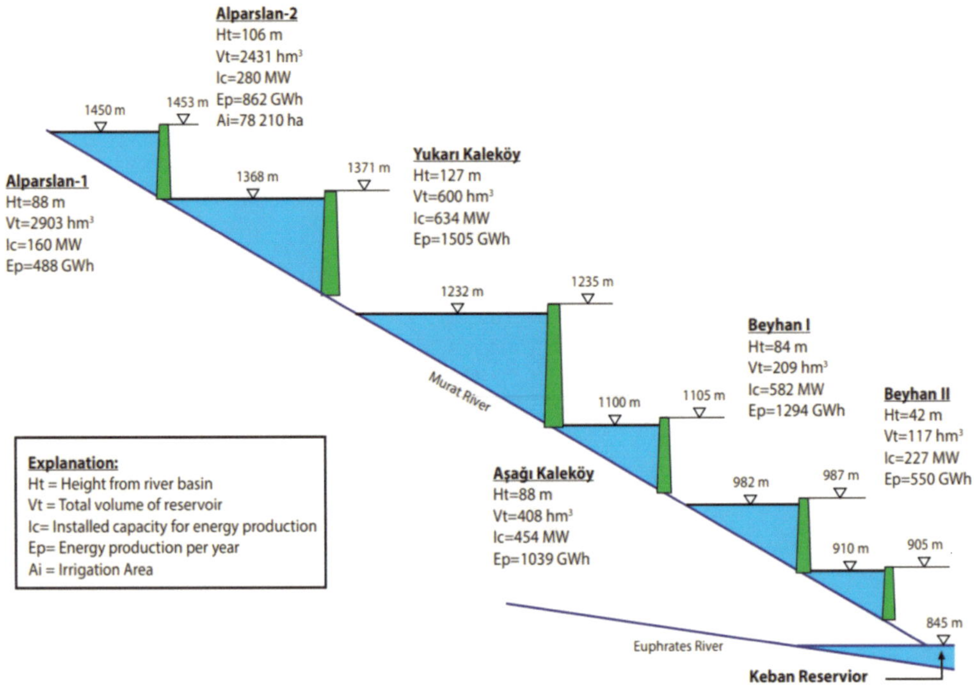

Figure 2.
The cascade system of Murat River of Euphrates-Tigris Basin [13].

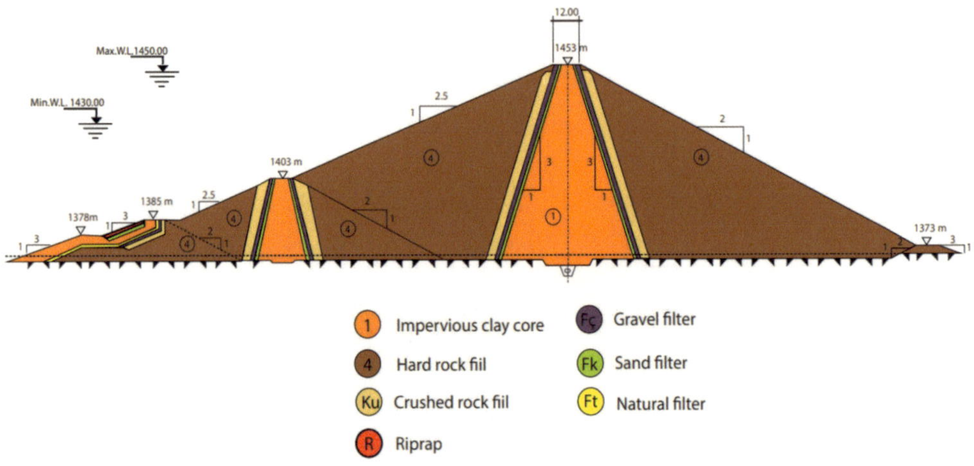

Figure 3.
The cross section of the Alparslan-1 dam [13].

evaluated in detail below. Alparslan-1 and Lower Kalekoy dams, which are essential structures for the cascading system, are also included in this evaluation.

The Alpaslan-1 dam, which is the upmost structure in the cascade system of Murat River, is the most critical element according to failure scenarios. It was designed as the rock-fill type with an embankment body volume of 3.8 million m^3 (**Figure 3**). Alparslan-1 dam has a height of 88 m from the river bed and a crest length of 571 m. Its total storage and active capacities are 2993 and 1743 hm^3, respectively. The dam is under the impact of the near energy source. In other words, it is 8.9 km far away from the energy source, which can produce an earthquake of magnitude 6.4. The dam site poses a very high risk for all dams of the cascading system. This means that the overtopping problem will be seen in all the dams located downstream of the Alparslan-1 dam. The partial failure analyses indicate that a minor breach forms through the dam's embankment. In complete failure scenarios of Alparslan-1, it is expected to have a large breach form in the embankment of the Alparslan-2 dam. In this context, considering the type of embankment material and foundation as well as the stability of valley slopes, a detailed geotechnical study should be carried out within the scope of the relevant dam safety program, and failure analyses should be conducted under different loading conditions using actual data for the foundation and embankment materials.

Alparslan-2 dam, located within the cascade system and at the downstream side of Alparslan-1 Dam, has a significant storage capacity (2431 hm^3). However, its storage capacity is smaller than that of the Alparslan-1 dam. This dam was designed as a rockfill with an asphaltic core and the highest embankment on the Murat river (height from river bed = 108 m) with an embankment volume of 12.45 million m^3. The dam site is not affected by the near-source zone, and the PGA values that will occur are relatively low. The dam site is 13.2 km away from the energy source zone, which could produce a magnitude 6.4 earthquake in Mw. The dam's most significant source of hazard is large-scale damage or failure at the Alparslan-1 dam located upstream. Rather than measures to increase stability in the dam, construction measures should be taken to discharge the flooding resulting from the Alparslan-1 dam. For this purpose, the discharge capacity of the spillway can be increased,
and/or additional spillway discharge capacity is created. This kind of study will be conducted for the first time in the country. Therefore, detailed evaluations should be made within the framework of the national dam safety program. It is a well-known fact that there can be some drawbacks to the private sector's owning a dam for a certain period of time. This is an essential factor that may effectively take additional construction measures that may be necessary to be done in the project.

Figure 4.
A general view from the Upper Kalekoy Dam.

The Upper Kalekoy dam is the highest dam in the cascade system of Murat River (127 m in height from the river bed) and was designed as the type of roller-compacted concrete dam (**Figure 4**). Its maximum storage volume is 784 hm3, and there are three separate dams (Aşağı Kalekoy, Beyhan-I, and Beyhan-II) built with a lower capacity and a rigid type on its downstream. The dam site has the highest earthquake hazard in the cascade system, which is 4.3 km away from the energy source zone, capable of producing a magnitude of 6.1 earthquakes in Mw. The dam body will be able to work as a submerged structure during an overtopping event resulting from a large flood flow that may occur at the upstream.

The Lower Kalekoy dam, which has a height of 88 m from the river bed and a storage capacity of 517 hm3 at maximum water level, is built as a concrete-type downstream of the Upper Kalekoy dam. The dam site is located 12.7 km away from an energy source zone that could produce an earthquake with a magnitude 6.4 in Mw. Therefore, it has a lower risk of exposure to peak ground acceleration (PGA) than others located upstream and downstream. The dam body can be operated as a submerged structure during an overtopping event.

Beyhan-I dam, having a height of 84 m measured from the river bed and a storage of 369 hm^3 at the maximum water level, is located between Lower Kalekoy and Beyhan-II dams and designed as a roller-compacted concrete dam. The dam site is under the impact of the near-field motion. The dam location is 3.0 km away from the fault zone, which could produce an earthquake with a magnitude of 6.3 in Mw. It would be subjected to the high PGA value (0.551 g) that may occur from the respective source. The dam body will be able to act as an overall spillway during considerable flooding by dam failure on its upstream.

Beyhan-II dam is the structure with the lowest height from the river bed (42 m) and the lowest storage capacity (79 hm3) on the cascade system of Murat River. The dam site is also the closest to the energy source (1.7 km), which can produce an earthquake with a magnitude of 6.3. Therefore, the dam site's largest PGA values were obtained in the deterministic and probabilistic seismic hazard analyses conducted within the scope of this study. This dam body, a built-in concrete gravity type, will be able to perform as a submerged structure while an overtopping event occurs during considerable flooding resulting from dam failure downstream.

The cascade system with six large dams on the Murat River merges with the Keban dam reservoir located on the main Euphrates river. Keban dam is the second largest storage dam in the country, with a storage capacity of 31,000 hm3. The largest catastrophic failure that may occur in the cascade system of Murat River can be met by the huge capacity of the Keban dam when operated with an appropriate program. However, it is necessary to lower the normal water level (NWL) of the Keban dam reservoir by 3 m in terms of the system's reliability.

5. Conclusions

The reliability of a cascade dam system depends on the type and storage capacity of each dam structure included in the system. In other words, the total risk of a dam located in a cascade structure and a singular dam located on a river is not the same. The total risk classifications used for this purpose do not take into account this difference and even suggest lower risk values related to the evacuation requirement and damage rate that may occur downstream. The hazard can be more due to the triggering phenomena for a significant catastrophic failure on the cascade system upstream. In reality, the total risk is several times greater than that estimated by risk classification.

Among the dams located in the Murat river cascade system, the maximal total risk belongs to the Alparslan-1 and Alparslan-2 dams, built in the embankment type and located at the most upstream part of the system. All design and construction measures to enhance safety should be taken for the Alparslan-1 dam, and the necessary observations and measurements should be carried out sensitively. The author introduces that a specific operation program should be prepared for this dam reservoir and operated by experienced engineers and technicians. For the Alparslan-2 dam, which is of the rock-fill type with a central asphalt core, construction measures should be taken to prevent overtopping phenomena and/or reduce its impacts. One of the considerations is to increase the discharge capacities of the existing dam system by using additional spillways/spillways to be constructed during the operation stage.

The most important advantage of the cascade system is that rigid dams can work as an overall spillway during overtopping. However, this issue is not considered in assessing the total risk analysis methods. It is appropriate to make a specific renovation to the cascade system in these methods. All of the dams involved in the study, except two separate dams mentioned above, are built in a rigid dam type, and the entire body can operate as a spillway during significant flooding. This is one of the most important advantages of the cascade system of Murat River. Another important advantage is a Keban dam with a huge storage capacity located downstream of the system. However, it would be appropriate to withdraw the normal water level to the elevation of 842 m for the Keban dam. This will also improve the safety of the dam, which is 48 years old and is of the composite type.

Author details

Hasan Tosun
Geotechnical Engineer Rector of Mudanya University, Bursa, Turkey

*Address all correspondence to: hasan.tosun@mudanya.edu.tr

References

[1] ICOLD. Selecting Parameters for Large Dams-Guidelines and Recommendations. ICOLD Committee on Seismic Aspects of Large Dams. Bulletin 72. 1989

[2] Bureau GJ. In: Chenh WF, Scawthorn C, editors. Dams and Appurtenant Facilities in Earthquake Engineering Handbook. Bora Raton: CRS Press; 2003. pp. 26.1-26.47

[3] ICOLD. Selecting Seismic Parameters for Large Dams-Guidelines. ICOLD, Bulletin 148, 2016

[4] Tosun H. Seismic studies. International Water Power & Dam Construction. 2006;**58**(2):20-23

[5] Tosun H, Zorluer I, Orhan A, Seyrek E, Savaş H, Türköz M. Seismic hazard and total risk analyses for large dams in Euphrates basin, Turkey. Engineering Geology. 2007;**89**(1-2):155-170

[6] Tosun H. Total risk analysis of dam and appurtenant structures in a basin and a case study. International Congress in River Basin Management. 2007;**I**:477-488

[7] Tosun H, Türkoz M, Savas H. River basin risk analysis. Int. Water Power and Dam Construction. May issue; 2007

[8] Tosun H. Evaluating earthquake safety for large dams in Southeast Turkey. Hydro Review Worldwide (HRW). 2008:34-40

[9] Tosun H. Re-analysis of Ataturk Dam under Ground Shaking by Finite Element Models. In: Proceeding of CDA Annual Conference. Saskatoon, Canada. 2011

[10] Tosun H. Earthquake Safety of Keban Dam, Turkey. In: Proceeding of CDA Annual Conference. Fredericton, NB, Canada; 2012

[11] Tosun H, Oguz S. Stability Analysis of Atatürk dam, Turkey as Based on the Updated Seismic Data and Design Code. In: Proceeding of 85th Annual Meeting of International Commission on Large Dams, Prague, 66-66, 2017

[12] Tosun H and Oguz S. Seismic Design for Existing Large Dams and Case Studies. In: Proceeding of 11th ICOLD European Club Symposium Chania, Crete. 2019

[13] DSI. Dams of Turkey. TR-COLD, Ankara, 2016602 p

[14] Fraser WA, Howard JK. Guidelines for Use of the Consequence-Hazard matrix and Selection of Ground Motion Parameters. Technical Publication, Department of Water Resources. Division of Safety of Dams. 2002

[15] Jiminez MJ, Giardini D, Grünthal G. Unified Seismic Hazard Modelling throughout the Mediterranean Region. Bolettino di Geofisica Teorica ed Applicata. 2001;**42**(1-2):3-18

[16] Kramer SL. Geotechnical Earthquake Engineering. Upper Saddle River. NJ 653 p: Prentice-Hall; 1996

[17] Krinitzsky E. Discussion on Problems in the Application of the SSHAC Probability Method for Assessing Earthquake Hazards at Swiss nuclear power plants. Engineering Geology. 2005;**82**:62-68

[18] Tosun H, Seyrek E. Total risk analyses for large dams in Kizilirmak basin, Turkey. Natural Hazards and Earth System Sciences. 2010;**10**(5):979

[19] Seyrek E, Tosun H. Deterministic approach to the seismic hazard of dam sites in Kızılırmak basin, Turkey. Natural Hazards. 2011;**59**(2):787

[20] Seyrek E, Tosun H. Influence of analysis methods for seismic hazard on

total risk of large concrete dams in Turkey. *Gazi Univ,* J.Fac Engineering Architecture. 2013;**28**:67-75

[21] Tosun H. Earthquakes and dams. In: Moustafa A, editor. Earthquake Engineering—From Engineering Seismology to Optimal Seismic Design of Engineering Structures. London: IntechOpen; 2015. DOI: 10.5772/59372

[22] Tosun VT, Tosun H. Total Risk and seismic hazard analyses of large dams in Northwest Anatolia, Turkey. In: Proceeding of 85th Annual Meeting of International Commission on Large Dams. Prague. 2017. pp. 165-165

[23] Tosun H. Safety assessment of large reservoir constructed for domestic water near urban areas and a case study. In: Proceeding of ICOLD-ATCOLD Symposium on Hydro Engineering, Wien. 2018. pp. 917-927

[24] Tosun H. Stability analysis of large embankment dams located on shear zones and case studies. Hydrovision International. Colorado Convention Center. 2017

[25] Tosun H, Tosun VT. dynamic analysis of embankment dams under strong seismic excitation and a case study. In: Proceeding of Long-Term Behaviour and Environmentally Friendly Rehabilitation Technologies of Dams (LTBD 2017). 2017. DOI: 10.3217/978-3-85125-564-5-102

[26] Tosun H. Seismic stability of large dams located near energy source and a case study. In: Proceeding of 5th International Conference on Earthquake Engineering and Seismology, METU-Ankara. 2019

[27] Tosun H. Earthquake safety evaluation for large dams located near the energy source and case studies. In: Proceeding of 11th ICOLD Europen Club Symposium. Chania. 2019

[28] Tosun H. Hazard and total risk analyses of large dams under threat of the north anatolian fault zone in Mid-Anatolia, Turkey. In: Proceeding of 5th World Congress on Civil, Structural, and Environmental Engineering (CSEE'20) Virtual Conference. 2020. DOI: 10.11159/icgre20.191

[29] Tosun H, Tosun VT, Hariri-Ardebili MA. Total risk and seismic hazard analysis of large embankment dams: Case study of Northwest Anatolia, Turkey. Life Cycle Reliability and Safety Engineering. 2020;**1-10**. DOI: 10.1007/s41872-020-00113-4

[30] Tosun H. Experience on earthquake safety of large embankment dams constructed in Turkey. In: Proceedings of the 6rd International Conference on Civil Structural and Transportation Engineering (ICCSTE '21). Virtual Conference. 2021. pp. 1-8 (addition). DOI: 10.11159/iccste21.137

[31] Tosun H. Total risk and seismic stability of existing large CFRD's in Turkey. In: Proceedings of the 6rd International Conference on Civil Structural and Transportation Engineering (ICCSTE '21) Virtual Conference. 2021. DOI: 10.11159/iccste21.138

[32] Tosun H. Dynamic stability of large embankment dams and a case study. In: Proceedings of the 6rd International Conference on Civil Structural and Transportation Engineering (ICCSTE '21). Virtual Conference. 2021. DOI: 10.11159/iccste21.139

[33] FEMA. Federal Guidelines for Dam Safety—Earthquake Analyses and Design of Dams. 2005

[34] DSI. Selection of Seismic Parameters for Dam Design. State Hydraulic Works. Ankara. 29 p, 2012 (in Turkish)

Chapter 2

Geomembranes in Dam Engineering

Gabriella Vaschetti

Abstract

Geomembrane systems are used to provide, enhance, or restore watertightness in dams since 1959. In new construction, they are installed on embankment dams, RCC dams, and cofferdams, while in rehabilitation they are used on all types of dams. They can be installed as a full-face liner, or to line parts of the dam where a higher risk of infiltration is expected, or as external water stop at peripheral and vertical joints and at contraction joints. They can be exposed to the water of the reservoir or be covered by a ballast layer; a watertight seal at all peripheries prevents water infiltration underneath the geomembrane liner. A geomembrane water barrier is a technically and cost-effective sustainable solution. The chapter discusses the design of the state-of-the-art solutions, the technical and economic advantages, installation aspects, performance, and references, with significant examples of all available options. A recent solution for underwater placement, developed for repair but applicable also in new construction, will be presented.

Keywords: geomembrane, geocomposite, watertightness, joints, underwater

1. Introduction

Preventing or minimising water infiltration from the reservoir is crucial for the behavior of a dam because, over time, persistent seepage can make the dam deviate from its design conditions, ultimately jeopardizing its integrity. To ensure safe operation over their service life, dams are built to be intrinsically watertight or to have an upstream or an internal water barrier.

In 1959, synthetic watertight materials were first adopted in dams to substitute traditional water barriers such as concrete or clay. With a few exceptions (geotextiles impregnated in-situ with bitumen), these synthetic materials are factory-produced in form of thin continuous flexible sheets. In 1977 Dr J. P. Giroud proposed to name them "geomembranes" [1], which is the terminology internationally adopted ever since. When installed in dams, the thickness of geomembranes is in the range of 2 to 3.5–4 mm.

Several types of geomembranes are available, with different characteristics and different behavior in service, depending on the components, the formulation, and on the type of reinforcement if any. The International Commission on Large Dams (ICOLD) has extensively addressed these issues [2], also giving information on the number, type, and years of installation for the different types of geomembranes adopted in dams. Plasticised polyvinylchloride (PVC) geomembranes resulted to be by far the ones most used in dams. In the majority of applications, the PVC geomembrane is heat-bonded at fabrication to an anti-puncture geotextile, to form a "composite geomembrane", widely albeit inappropriately known as "geocomposite". Although the term "geocomposite" is a general term indicating a geosynthetic composed of two different materials, not necessarily including a geomembrane, within this chapter the term "geocomposite" will indicate a composite geomembrane.

PVC geocomposites and geomembranes are preferred among other types of geomembranes for dams due to their favorable tensile behavior and puncture resistance. In particular, when it comes to resistance to differential settlements, a study by Giroud and Soderman [3] has shown that an appropriate combination of tensile strength and strain is essential. This optimum combination depends on the shape of the tension-strain curve of the geomembrane, which should show a monotone behavior with possibly no yielding peak, a low modulus, and a large elongation capacity. The suitability of a geomembrane to resist differential settlements can be quantified in terms of Co-Energy, a rationale defined by Giroud [3, 4]. The Co-Energy represents the ability of a geomembrane to withstand a combination of stress and elongation, which can be representative, for instance, of an in-field occurrence of differential settlements. The Co-Energy is determined by the area between the tension-strain curve and the tension axis (i.e., the vertical axis) and expressed as energy per unit area of the geomembrane. The larger the area, the higher the capability to resist differential settlements. **Figure 1** shows the definition of the Co-Energy in the monotone increasing portion of the tension-strain curve of two considered geomembranes: a 3 mm thick SIBELON® geocomposite (SIBELON® PVC geomembrane heat-bonded to a non-woven geotextile), up to the geotextile break tension, and a 3 mm thick High-Density Polyethylene (HDPE) geomembrane, up to the characteristic yielding peak.

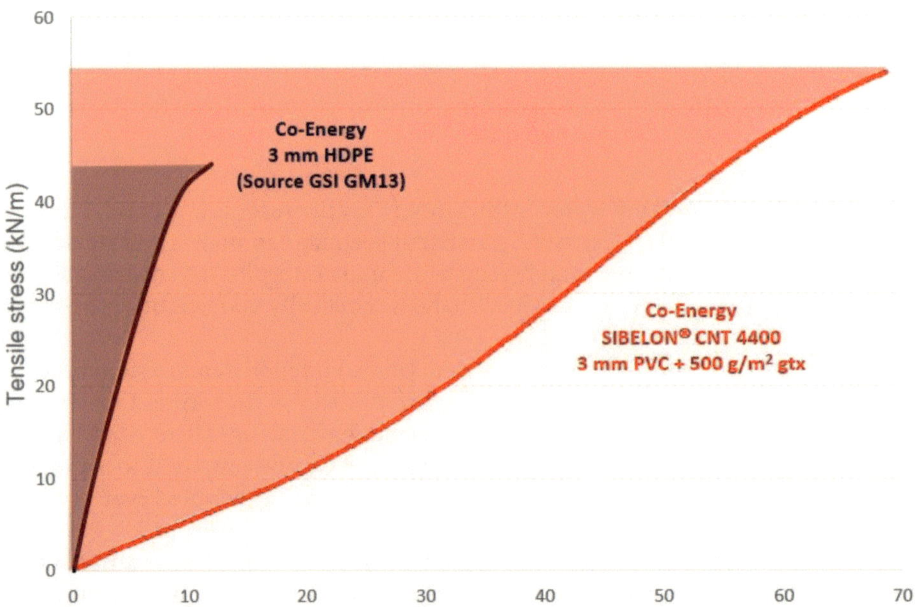

Figure 1.
Comparison between co-energy of a 3 mm thick HDPE geomembrane—the darker area—and of the geocomposite SIBELON® CNT 4400 (3 mm thick SIBELON® geomembrane heat bonded during fabrication to a 500 g/m² nonwoven geotextile)—the red area.

Figure 1 clearly shows that the maximum allowable Co-Energy associated with the SIBELON® geocomposite is significantly greater than the Co-Energy associated with the HDPE geomembrane. As a result, the factor of safety with respect to a potential differential settlement is significantly higher for a SIBELON® geocomposite than for an HDPE geomembrane. It is important to note that a different geomembrane thickness would not produce a different result.

Successful performance of a geomembrane system depends on the physical and mechanical characteristics of the geomembrane, but also on the design and on the quality of the installation. A material with potentially optimal characteristics will still perform badly if the geomembrane system is not properly designed and installed. Bulletin 135 [2], provides guidelines for the design of geomembrane systems in different types of applications, and makes recommendations for Quality Control, installation procedures, specifications, and contracts, with the objective of ensuring long-lasting efficient water barrier. The sections that follow give an up-to-date overview of the various types of applications in new construction and in rehabilitation.

2. Design considerations

The design of a geomembrane system is closely connected to the characteristics of the dam and of its environment. In addition to the specific competencies related to geomembranes, the designer of the geomembrane system should have a thorough knowledge and understanding of the behavior of the dam and of the loads that will be acting on the geomembrane and possibly work in cooperation with the designer of the dam in case of new construction, and with the owner in case of rehabilitation.

"Environmental" loads, such as ultra-violet radiation, chemical attack by substances diluted in the water of the reservoir or by flora and fauna, and vandalism, influence the selection of the geomembrane type, thickness, and reinforcement, and to a certain extent the choice of whether to leave the geomembrane exposed or to cover it. The mechanical loads are the basis for the calculation of the anchorage system, which in turn is influenced by the type and thickness of the selected geomembrane.

2.1 Exposed geomembranes

The design of exposed geomembrane systems is based on the concept of keeping the geomembrane stable, taut to the dam face to avoid an insurgence of wrinkles and folds where stress concentration can occur, and capable of sustaining the applied loads. The static load of the water in the reservoir, associated with the conditions of the surface on which the geomembrane is laid, and the gravity load due to the weight of the geomembrane itself, will steer the selection of a material capable to resist puncture, burst, and tensile stresses. The dynamic loads due to uplift by wind when the reservoir is empty or lowered, by waves, differential deformations, and by earthquakes, are the parameters for designing the face anchorage system and the perimeter seals.

In the majority of cases, the worst-case scenario is uplift by wind, which will cause a deformation of the geomembrane and resulting stress on the face anchorage system. The equations developed for the face anchorage system [5–7] are based on the scheme shown in **Figure 2**.

Based on the scheme of **Figure 2**, being L the length of the undeformed geomembrane (Geomembrane before uplift) between adjacent anchorage points A and B, and 2θ the angle of the arc of the deformed geomembrane (Geomembrane after uplift), the

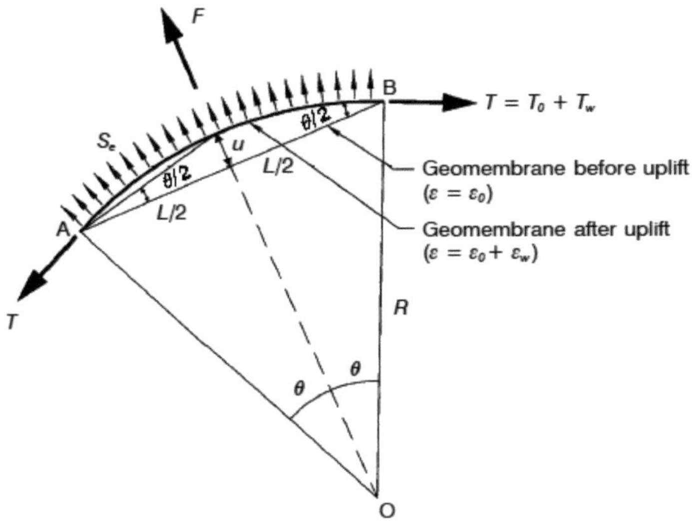

Figure 2.
Schematic representation of an uplifted geomembrane (from Giroud, 1995 and 1997).

equations are used to estimate, depending on the properties of the geomembrane used, the force S_e applied to a unit width of lining between the two adjacent anchorage points, the displacement and subsequent strain T experienced by the geomembrane lining, and the resulting perpendicular and tangential stresses applied to the anchorage points.

When designing face anchorage systems for dams located in highly seismic areas, the worst-case scenario can be the maximum expected seismic event causing opening/widening of a crack when the reservoir is full, and the water pressure will force the geomembrane from its initial "supported" configuration into a deformed configuration, which must be verified to ensure that the strains in the geomembrane stay within the acceptable limits. An approach used by geomembrane systems designers assimilates the phenomenon of deformation due to water pressure to the phenomenon of deformation due to wind uplift. This same approach is used to verify the capability of external water stops to resist the displacements at joints between monolith blocks in RCC dams, and at peripheral joints and vertical joints between face slabs in Concrete Face Rockfill Dams (CFRDs).

Another critical component is the perimeter seal, which must maintain its watertightness against the applied water head, under any operating condition. Means to verify the watertightness of tie-down seals in static conditions in the laboratory and in the field are available. In case the area of the seal is susceptible to differential settlements (i.e., in dynamic conditions) since at present a test to verify watertightness in dynamic conditions has not yet been developed, design must use the available calculation methods to ensure that the seal is compatible with the expected event and that the geomembrane can bridge the differential settlement without sustaining the initial deformations and tensions induced by it, but only the deformations and tensions due to the hydrostatic pressure. To analyse at best which geomembrane can better perform in this respect, the concept of Co-Energy can be used, as mentioned in section 1.

Perimeter seals of the embedded type require a different approach: the watertightness of the seal cannot be verified nor calculated and shall rely on the engineering judgment of the designer for dimensioning the seal, and on the experience, craftsmanship, and Quality Control procedures of the installer for executing a correct seal. To achieve a watertight barrier, in case of a trench filled with low permeability material (e.g., clay) it will be necessary to correctly execute excavation, filling, and compaction of the trench, while in case of a slot filled with watertight resin, testing for pull-out, and control of correct filling of the slot and correct polymerisation of the resin, will be needed.

2.2 Covered geomembranes

Totally covered geomembranes are theoretically protected from environmental loads, with the possible exception of chemical attacks. For partially covered geomembranes, generally adopted for aesthetical reasons or to protect against fall of rocks at the abutments or vandalism in accessible areas, a face anchorage system may be needed where the geomembrane is exposed to wind action, depending on the extent of the exposed area and on the wind speed.

The type and thickness of the cover layer, different depending on the type of dam, are discussed in section 3. The design shall consider the uplift actions, and, in the case of embankment dams, the puncture loads exerted by the cover layer and the stability at sliding of the overall cover system, as further discussed. For the perimeter seals, the considerations made for the exposed systems generally apply.

3. Applications

Sections 3.1 to 3.3 address the waterproofing of the various categories of dams. Since the concepts and components are essentially the same, what is discussed here for full-face waterproofing applies also to waterproofing of parts of the dam where the leakage is expected/experienced. Section 3.4 discusses waterproofing of joints, and section 3.5 a special solution for underwater placement.

Geomembrane systems are the most sustainable waterproofing technology for dams: for all types of applications, the components have small volume and lightweight, and their installation does not require heavy equipment, so no large areas are required for site organisation, construction, or processing plants for materials. The environment and the communities will not be affected by excavations or by heavy transport, and transport to remote sites not accessible by vehicles will be feasible even by helicopter at sustainable costs. Since installation is quick and can be carried out in almost any weather condition, the project will be completed, the dam put in operation, and the site restored to its foreseen final conditions, or to its initial conditions in case of rehabilitation, in shorter times.

3.1 RCC dams

In new RCC dams, geomembranes are adopted as a water barrier for dams using an RCC mix with low cementitious content, assuming that the RCC provides the stability and the geomembrane placed at the upstream face permanently grants watertightness to the bulk RCC material, to the lift joints and construction joints, and to any thermal cracks or cracks that may develop due to other causes. With this philosophy of separating the static function from the waterproofing function, the cementitious content can be reduced, pozzolan and fly ash even eliminated, and requirements for placement and quality control of the RCC body can in general be relaxed. Lower cement content results in lower production of hydration heat, therefore the need to control the temperature with expensive cooling devices is reduced. GEVC (Grout Enriched

Vibrated Concrete) or GE-RCC (Grout Enriched RCC) or CVC (Conventional Vibrated Concrete) at the upstream face can be deleted from the design, and horizontal joint treatment or the bedding mix on lift surface areas can be reduced. It is thus possible to place one RCC mix over the entire cross-section of the dam without the interferences caused by placement of bedding mix/CVC. The treatment of the lift joints may still be needed but only for shear strength and dam stability. Vertical waterstops and drains can also be eliminated because they are included in the geomembrane waterproofing and full-face drainage system. Local material more easily available and aggregates of less stringent properties can be used for the RCC mix, and less stringent construction and QA/QC procedures can be implemented. The overall result is that construction costs and construction time are remarkably reduced, and quicker filling and operation of the facility can be achieved.

The geomembrane can be installed on the completed upstream face and left in contact with the water of the reservoir (exposed geomembrane, first project in 1990), or the geomembrane can be embedded in precast concrete panels used as permanent formworks to build the dam (covered geomembrane sandwiched between the concrete of the panels and the RCC, first application in 1984).

3.1.1 Exposed geomembrane

The exposed configuration is the most frequently adopted worldwide and has been installed on very high RCC dams like Miel I (188 m, Colombia), Balambano (99.5 m, Indonesia), Olivenhain (97 m, USA), and Susu (90 m, Malaysia). The geomembrane liner must be kept stable and taut on the upstream face to avoid it being uplifted by wind and waves, and to minimise folds and wrinkles where stress concentration can occur. The state-of-the-art face anchorage system, discussed also in Bulletin 135 [2], consists of two stainless-steel profiles, the first one in the shape of a U that is secured to the upstream face, and the second one in a shape similar to the Greek letter Omega, which is installed at the overlapping of two adjacent vertical geomembrane sheets, fastened to the underlying U profile, and waterproofed by a geomembrane cover strip heat-seamed on the liner. In older projects, the U profile was embedded in the RCC lifts as they were being placed, so that one component of the anchorage system was already in place when the dam body had been completed. However, this method presented some drawbacks, and it was gradually substituted: since 2007, the U profiles are generally installed after RCC placement has been completed. The case history that follows discusses the typical details for face anchorage, peripheral anchorage, and drainage and monitoring systems of a recent project on a high RCC dam.

Susu dam in Malaysia is part of the Ulu Jelai Hydroelectric project commissioned by Tenaga Nasional Berhad (TNB), the largest power utility in Malaysia, for meeting the surge in demand for electricity at peak hours of the day. Susu RCC dam, 90 m high and about 512 m long at crest, was designed with an RCC mix of medium-low cementitious content (100 kg/m^3 of cement and 80 kg/m^3 of fly ash) and an upstream PVC geomembrane embedded in pre-cast concrete panels used as formworks to place the RCC (covered geomembrane system). This covered geomembrane solution had been used for the first time at Winchester dam, in Kentucky, in 1984. The major advantage of the covered geomembrane system is that the geomembrane is permanently shielded against potential environmental damage during its service life. However, the geomembrane is not shielded against the risk of damage during construction, which actually has been found to be the highest risk for the integrity of a geomembrane, and the exposed configuration has some additional advantages over the covered configuration:

its efficient face drainage system at the design stage allows reducing uplift, and during service, it allows removing infiltration and saturation water behind the waterproofing liner, so that saturation levels and pore pressures in the dam are lowered, with beneficial effects on uplift pressures, on safety factors, on AAR phenomena, and on appearance at the downstream face. The drainage system, through measurement of drained water, allows monitoring of the performance of the waterproofing system on a continuous basis and can be associated with other systems (Optical Fibre Cables, piezometers) to refine the assessment of the overall behavior of the geomembrane barrier. An exposed geomembrane liner can be inspected, controlled, and if needed repaired from accidental damages, even underwater, over its service life. Based on these considerations, the original design was modified to an exposed drained geomembrane system, which allowed reducing the cement content and removing the fly ash.

The geomembrane liner is a SIBELON®CNT 4400 composite geomembrane, formed by a 3 mm thick SIBELON® geomembrane heat-bonded at fabrication to a 500 g/m^2 non-woven needle-punched polypropylene geotextile. The geocomposite is fastened to the dam face along vertical lines of the well-known and afore discussed patented tensioning system of stainless-steel profiles, at regular 5.70 m spacing; like in all recent projects, the U profile is fastened by stainless-steel anchor rods embedded in chemical phials. The components of the drainage system are then installed: a high transmissivity drainage geonet is placed over the entire upstream face of the dam, and an additional band of geonet, 1 m high, is placed along the bottom perimeter of each of the three horizontal drainage sections, to act as drainage collector. Drainage discharge pipes and ventilation pipes are constructed to reach the two drainage galleries at elevations 470 m a.s.l. and 518 m a.s.l., and at crest. In total 10 drainage compartments, 13 drainage pipes, and 30 ventilation pipes maintain the drainage geonet at atmospheric pressure. The waterproofing liner is placed over the geonet and secured with the Omega profiles, as shown in **Figure 3**.

Figure 3.
Susu RCC dam. From right to left, the vertical U profiles fastened to the dam, the black drainage geonet placed between them on the upstream face, the waterproofing liner being placed on the geonet, some Omega profiles providing temporary anchorage against wind uplift, and at the extreme left the Omega profiles permanently fastened.

The drainage system constitutes Susu the monitoring system for the behavior of the geomembrane liner.

The geomembrane system, which extends from crest at elevation 547.90 m a.s.l. down to the grouting plinth, is divided into three horizontal sections terminating at elevation 519.00 m a.s.l., at elevation 471.00 m a.s.l., and at the intersection with the grouting plinth, respectively. In the section below elevation 474 m a.s.l., which is covered with backfill, the waterproofing geocomposite was protected with a double layer of geotextile for preventing damage during placement of the backfill.

The geomembrane liner must be sealed at peripheries by a watertight anchorage preventing water infiltration under the liner. The seal is watertight against water in pressure around the submersible perimeter, and against rain waves and snowmelt at crest. In RCC dams, submersible seals are of the tie-down type and made on conventional concrete: typically, they consist of a stainless-steel batten strip fastened with chemical anchors at 150 mm spacing, rubber gaskets and epoxy resin between the batten strips and the subgrade, and stainless-steel splice plates at abutting batten strips - at Susu, they are 80x8 mm in section and were placed along the spillway, the grouting plinth, the diversion conduits wall, and around the trash rack. At crest, the seals are more flexible stainless-steel batten strips, 50x3 mm, fastened with expansion anchors at 200 mm spacing, with a neoprene gasket placed between the subgrade and the waterproofing geocomposite - at Susu, they were placed at crest out of the spillway section and at the two junctions between adjacent horizontal sections, where watertightness was provided with a cover strip of a 3.00 mm thick SIBELON® C 3900 geomembrane, the same used to waterproof the Omega profiles.

Placement of the RCC, for a total of 731,000 m^3, started in March 2014 and was completed in September 2015. Installation of the geomembrane system started in June 2015 and was completed in January 2016. Completion of the geomembrane system, for a total of 27,510 m^2, was longer than usual because the installation was planned to match the construction schedule of the dam, therefore performed in three separate phases: in the first phase, prior to starting of construction of the dam body, the grouting plinth was waterproofed with a geocomposite that was later watertight connected to the geocomposite lining the upstream face, so to create a continuous impervious barrier down to the grout curtain; in the second phase, once construction of the dam body had been completed, the geomembrane system was installed on the dam face; finally, in the third phase installation of the geomembrane system was completed

Figure 4.
View of the upstream face of Susu RCC dam at completion of the geomembrane system at spillway (left), and at start of impoundment (right).

at the spillway, upon construction of the concrete ogee crest (**Figure 4** at left). The reservoir was impounded shortly after the waterproofing works were completed.

3.1.2 Covered geomembranes

The covered configuration is used mainly in the USA where it was first conceived. During fabrication of the precast concrete panels that will be used as permanent formworks to place the RCC, the waterproofing geocomposite is placed on the fresh concrete, the geotextile side in contact with the panel, and vibrated (**Figure 5** at left) to make the concrete impregnate the geotextile and thus attach the geocomposite to the panel. The geotextile also has an anti-friction function in case of differential movements, when if stresses are high it can detach from the geomembrane layer avoiding possible damage to the water barrier.

After the first row of starter panels, watertight connected to the plinth, has been placed, a few rows of formwork panels are erected, and their junctions are waterproofed by strips of geomembrane of the same type used for the panels. A protection geotextile is then placed on the geocomposite, and the spreading and compacting of the RCC + upstream additional layers (GEVC, COV, or other) if any starts.

An interesting example, with a mixed configuration to the author's knowledge adopted for the first time ever, is Rizzanese, a 40.5 m high RCC dam in the island of Corsica owned by EDF and used for hydropower. The dam has a central curved spillway section, made in RCC up to elevation 534 m, and in CVC at spillway and chute, and two lateral sections made entirely in RCC up to crest at elevation 546.5 m. The upstream face is inclined 1H/1V in the bottom part up to elevation 520 m, and vertical up to crest. EDF designed an RCC mix with 80 kg/m^3 of cement, and a PVC geocomposite as an upstream water barrier on the entire upstream face. At the inclined section under elevation 520 m, the upstream face was made with prefabricated concrete elements, on which a geocomposite SIBELON® CNT 3750 (2.5 mm thick SIBELON® geomembrane heat-bonded at fabrication to a 500 g/m^2 non-woven needle-punched polypropylene geotextile), was placed and covered by backfill placed on a 1000 g/m^2 anti-puncture geotextile. At the vertical section above elevation 520 m, the upstream face was formed by the precast concrete panels discussed before, embedding a SIBELON® CNT 2800 geocomposite (2 mm thick SIBELON®geomembrane + a 200 g/m^2 non-woven needle-punched polypropylene geotextile).

Figure 5.
Vibrating the concrete to attach the geocomposite to the precast panel (left), and Rizzanese RCC dam seen from the upstream: the bottom inclined section has not yet been lined, placement of RCC against the panels with embedded geocomposite + protection geotextile at the vertical section is ongoing (right).

A 30 cm thick layer of CVC was placed between the panels and the RCC. At the spillway, the waterproofing system is prolonged to waterproof part of the vertical sides and the spillway sill, where it is covered by a sacrificial geomembrane to protect it against possible damage due to the heavy reinforcement and to concreting. The plinth has been waterproofed at the inclined section on its downstream and horizontal sides by SIBELON® C 3250, a 2.5 mm thick geomembrane, and at the vertical section only at the joints, with a Carpi patented external water stop system. Due to construction issues of the main contractor, the waterproofing system at the inclined section had to be installed after the vertical section had been completed. Waterproofing works were accordingly carried out in 2011 at the plinth and at the vertical section (**Figure 5** at right) and in 2012 at the inclined section (**Figure 6** at left).

In total, the waterproofing system includes 1162 m^2 at the inclined section and 2124 m^2 at the vertical section.

3.2 Embankment and hardfill dams

In new embankment dams, the design concepts are to substitute a rigid upstream water barrier (e.g., concrete), or a traditional core (e.g., clay), with a highly deformable geomembrane, to construct a Geomembrane Facing Rockfill Dam/a Geomembrane Facing Earthfill Dam (GFRD or GFED), or a dam with a geomembrane core. Due to their tensile properties, geomembranes can make possible projects that would not be feasible with other systems, due for example to their capability to accommodating deformations that would exceed the resistance of a rigid water Barrier, or resisting differential movements occurring between the deformable dam body and the rigid ancillary structures maintaining the watertightness of the joints dam/concrete appurtenant structures.

An upstream geomembrane can enormously simplify the design of an embankment dam project: the dam can be constructed on highly deformable foundations, zoning of the dam can be modified to some extent, the upstream face can be constructed steeper so that the volume of fill is lesser, the diversion tunnel can be shorter and smaller, multiple lines of waterstops (case of CFRDs) can be reduced or eliminated, connections to concrete structures can be designed to accommodate large differential movements. Also, a geomembrane is an asset when suitable materials are not available at the site at acceptable costs (case of clay core dams).

Figure 6.
Rizzanese dam. Placing the geocomposite at the inclined section before covering it with anti-puncture geotextile and backfill (left), and the impounded dam.

Geomembranes provide additional advantages in terms of reduction of construction times, constraints, and costs. For example, installation/construction of the reinforced concrete face slabs in a CFRD, and placement of copper and PVC waterstops, can have a considerable impact on the overall construction schedule. In dams with clay or bituminous concrete cores, since the construction of the dam body and of the core are strictly related, the constraints imposed by the weather conditions, or any disruption in the placement of the filter material, or in placement/compaction of the impervious core, will affect the overall rate of construction of the dam. On the contrary, installation of a geomembrane system is practically unaffected by weather and can be scheduled in function of the schedule of construction and operation of the dam: the geomembrane can be installed when the dam is completed, or in the lower completed part of the dam while construction of the fill is ongoing in the upper part. In case of floods during construction, the already waterproofed lower part of the dam will be a barrier against the flood, increasing the safety of the project.

3.2.1 Exposed geomembranes

With an exposed system, zoning is not strictly required, a single fill material can be used, the thickness of the drainage layer can generally be reduced, and, depending on the design of the dam, it can act also as a base/anchorage layer for the waterproofing geocomposite. To maintain the liner stable on the upstream face there are basically three available configurations for the face anchorage system, depending on if the dam is raised with extruded porous concrete curbs as a finishing layer (Itá method), or if a traditional embankment is constructed.

When the upstream finishing layer is made by extruded porous concrete curbs, the face anchorage system is constructed by embedding in the curbs, while they are being raised, geocomposite anchor strips of the same type constituting the waterproofing liner. An anchor strip fixed to one curb overlaps the anchor strip fixed to the preceding curb, and the overlapping strips are heat-seamed to form continuous anchor lines. When the embankment has been raised to the final or pre-established height (case of staged construction), the waterproofing geocomposite sheets are deployed over the anchor strips and heat-seamed to them, to form a continuous watertight lining (**Figure 7**).

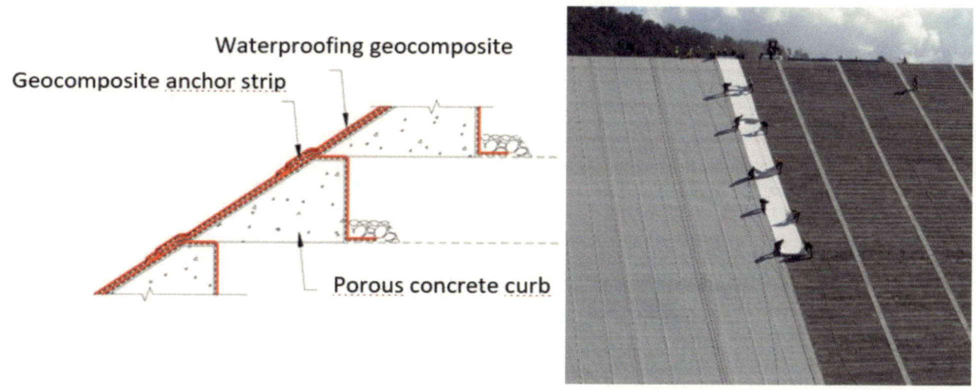

Figure 7.
Face anchorage with SIBELON® CNT geocomposite anchor strips embedded in curbs. At right the geocomposite being placed on the anchor strips at Nam Ou VI 88 m high rockfill dam in Laos.

This system has been applied in several projects, such as Nam Ou pictured below, Las Bambas GFRD in Peru (**Figure 8**), whose staged installation has so far reached 173 m, at the top part of Runcu 91 m high GFRD in Romania, at the top part of Rogun cofferdam in Tajikistan, whose staged installation has so far reached 80 m, and which will be incorporated in a 335 m high rockfill dam.

In dams where the finishing layer is not formed by extruded porous concrete curbs, depending on site-specific construction methods and type of supporting layers, the face anchorage system can be made by trenches or by deep anchors (**Figure 9**), positioned according to site-specific patterns that are a function of the uplift loads. If trenches are used, geocomposite anchor bands are embedded in trenches excavated in the compacted base layer, forming continuous anchorage lines. The trenches are generally backfilled with draining material (porous concrete or granular material), and the waterproofing geocomposite is heat-seamed to the anchor bands. The deep anchors face anchorage system consists of grouted anchors or earth "Duckbill" anchors that are driven into the embankment to provide the required pull-out resistance. The waterproofing geocomposite is anchored to the dam face by punching/clamping it over the anchors. A stainless-steel disk, an anti-puncture geotextile, and a geomembrane cover strip ensure watertightness where the anchors cross the waterproofing geocomposite.

The perimeter seals preventing water infiltration under the liner are as described for RCC dams when made on concrete. The bottom perimeter seal can also be made by embedding the waterproofing geocomposite in a trench that is then backfilled with

Figure 8.
Staged installation on curbs at Las Bambas dam, now reaching 173 m.

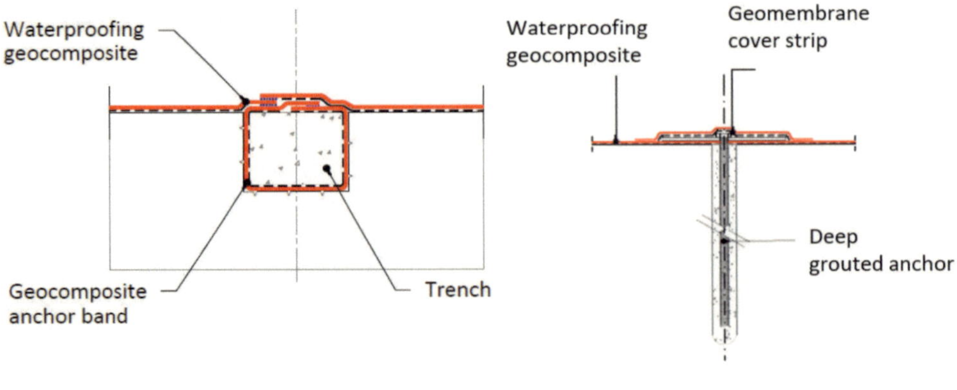

Figure 9.
Face anchorage by trenches and by deep grouted anchors.

low permeability material. The drainage system is site-specific and also in function of the construction procedure for the dam.

Examples of applications of anchorage by trenches are the bottom part of Runcu 91 m high GFRD in Romania, Murdhari 36 m high dam in Albania, Pico da Urze 31 m high dam in Portugal (**Figure 10**), Bulga 17 m high dam in Australia, and new reservoirs for pumped storage schemes in the Middle East and Morocco. Examples of applications of deep anchors are Filiatrinos 55 m high hardfill dam in Greece, Ambarau 21 m high hardfill dam in Congo and a section of the new Panama Canal 18 water-saving basins.

3.2.2 Covered geomembranes

In the covered system, the geomembrane can be placed at the upstream face and be covered with concrete or granular ballast that will provide the face anchorage, or it can be embedded in the dam as a zigzag central core or upstream inclined core. The design shall allow the waterproofing geomembrane to freely deform, minimising tensions due to movements in the subgrade on which the geomembrane rests. The puncture loads exerted by the cover layer, and the stability at sliding of the overall cover system shall be considered. The geomembrane is generally sandwiched between two geosynthetic layers that act as anti-friction layers and provide anti-puncture protection from the adjacent materials.

Examples of upstream covered systems are Bovilla 91 m high dam in Albania (**Figure 11** at left, unreinforced concrete cover), and Adret des Tuffes-Les Arcs 21 m high dam and reservoir in France (granular cover). A geomembrane core has been installed at Gibe III 50 m high rockfill cofferdam in Ethiopia, zigzag central geomembrane (**Figure 11** at right) that was preferred to a clay core and to a concrete or asphalt concrete facing respectively because of lack of local availability of material suitable for an impervious core, for sake of safety, simplicity (it would allow the realization of an embankment of homogeneous rockfill, with optimization in construction times and costs), and timing, as it would allow completing the construction within a very short construction period. An upstream inclined core waterproofs the bottom 65 m high section of Rogun cofferdam in Tajikistan.

Figure 10.
Installation of waterproofing liner over anchor trenches at Pico da Urze dam.

Figure 11.
At left Bovilla dam, at right Gibe III 50 m high cofferdam.

3.3 Concrete and masonry dams

In concrete and masonry dams, geomembranes are used to restore watertightness. They are always installed at the upstream face, in an exposed position unless site conditions require a partial cover layer. The state-of-the-art face anchorage system is the tensioning system described for exposed geomembranes in RCC dams. What is already discussed for these dams about perimeter sealing and drainage and monitoring systems applies also to these projects. Geomembrane systems that can be installed underwater are available for total and partial repair. They can be installed at any depth and provide long-term solutions when total or partial dewatering is not feasible or too costly.

3.4 Joints

Geomembrane systems can be conceived also for localised sealing, in form of "external waterstops", in new construction, and as a repair method.

In new construction, external waterstops are applied over the monolith joints of RCC dams and over the peripheral and vertical joints of CFRDs. The main technical advantage as compared to conventional embedded waterstops is that, thanks to a typical 40-50 cm width of waterproofing geocomposite free to slide and distribute stresses, they can bridge large openings providing permanent protection against localised leakage.

The design of external waterstops generally includes one or more geosynthetic layers, of different kinds and thicknesses depending on the extent of the movements expected by the designers of the dam and on the maximum water pressure. A typical external waterstop (**Figure 12**) comprises

- A single or multi-layered site-specific support/anti-puncture/drainage system, placed over the joint, in the area of the possible opening

- A waterproofing layer, typically the geocomposite already discussed in the previous sections, placed over the support layer, and anchored independently from it, to cover the joint, over an area exceeding by at least 200–250 mm on each side the area of possible opening of the joint

- A watertight perimeter seal of the tie-down type.

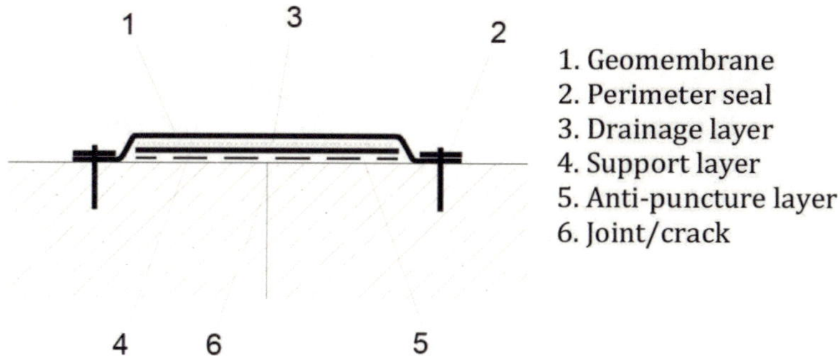

Figure 12.
Scheme of external waterstop (excerpt of ICOLD Bulletin 135).

The support/anti-puncture/drainage layers (**Figure 13**) avoid that the waterproofing geomembrane layer in contact with the water of the reservoir can be forced by the water pressure to intrude in the open joint. In dams located in seismic areas, the extent and frequency of the seismic cycles play an important role because they can cause an increase in water pressure that must be estimated and duly considered when designing the waterstop so that the admissible geomembrane elongation is not exceeded during and following the seismic event.

The most critical scenarios generally occur at the joints between a deformable embankment dam body and a rigid concrete appurtenance, or at construction joints of adjacent structures that may be subject to differential displacements, or in new CFRDs where the peripheral joint between the plinth and the face slabs is expected to have openings that will exceed the strength of the embedded waterstop, and therefore an additional waterproofing barrier is needed for safe operation. At such locations, it may be necessary to allow for an additional length of material so that the geomembrane liner can bridge the joint without sustaining the initial deformations and tensions induced by the opening of the joint, but only the deformations and tensions due to the hydrostatic pressure. The extra length can be created by placing under the waterproofing geomembrane liner a thick double-folded geotextile, or by inserting the external waterstop into a slot/recess formed at the joint. The extra length reduces the strain on the waterproofing geomembrane and ensures that the waterstop

Figure 13.
Detail and view of external waterstops installed at monolith joints of Platanovryssi 95 m high RCC dam in Greece.

deforms sufficiently to remain in contact with the surface on which it rests. In case the expected openings are very large, say in the order of tens of centimeters, the support layer is a high-tech textile (**Figure 14**) with a high modulus of elasticity and high tensile strength, which develops minimum elongations and high tensions, and thus restrains the deformation and consequently the tensions in the overlying waterproofing geomembrane.

The same concept (support/anti-puncture/drainage layers + waterproofing layer + watertight perimeter sealing) is used in the repair of failings joints and cracks, where conventional sealing methods, such as grouting and injections, often require recurrent interventions and put a strain on the maintenance costs and may not be adequate in case of large openings, which on the contrary geomembrane systems can accept.

Similar to other repair methods with geomembranes, also external waterstops have been adapted for underwater applications.

3.5 Special solution for underwater placement

A special solution for underwater placement, recently developed for repair of canals without stopping or reducing the water flow, is applicable also in new construction. This innovative system, the SIBELONMAT®, consists of two geomembranes connected to form a mattress, which is placed underwater covering the whole embankment to be watertightened, is temporarily anchored, and then filled with cement grout. The bottom geomembrane provides watertightness, the cement grout provides permanent anchorage by ballast, and the upper geomembrane provides containment for the fresh grout and enhances the hydraulic efficiency of the mattress due to its smoothness. The mattresses, prefabricated in panels of 10 m width and predefined length (**Figure 15** at left), are connected underwater by watertight heavy-duty zippers. The system has proven its efficiency in a few pioneer pilot projects in flowing water (**Figure 15** at right), which have shown that as more experience is acquired, improved design can be developed. Already under study are new components and equipment to improve and industrialise the system, with the objective of reducing installation times and costs. Such a system could be used for underwater lining of embankment dams and for blanketing, opening new scenarios that can potentially provide huge savings to the owners.

Figure 14.
High-tech support textile and waterproofing geocomposite at Angostura 32 m high 1.6 km long concrete face gravel dam, Chile. External waterstop on peripheral and vertical joints designed to resist joint openings up to 300 mm.

Figure 15.
The SIBELONMAT® panels under fabrication (at left) and injected and sealed at Kembs dike, part of the Grand Canal d'Alsace navigation canal in France.

4. Conclusions

Geomembrane systems are designed to allow the construction of safe dams, including dams of considerable height, with simple procedures that accommodate demanding scenarios and schedules, with sustainable environment-friendly solutions. They allow completing the waterproofing system for the dam in a shorter time and at lower costs than traditional water barriers. In form of external waterstops, they provide and maintain watertightness under loading conditions that would not be acceptable for embedded waterstops. They are a mature and reliable technology, as proven in hundreds of projects worldwide. They can be installed underwater, providing a long-term solution with no environmental impact.

Author details

Gabriella Vaschetti
Carpi Tech, Balerna, Switzerland

*Address all correspondence to: gabriella.vaschetti@carpitech.com

References

[1] Giroud, JP, Perfetti, J. Classification des textiles et mesure de leurs propriétés en vue de leur utilisation en géotechnique. In: Proceedings of the International Conference on the Use of Fabrics in Geotechnics; Paris; 1977. p. 345-352

[2] International Commission on Large Dams. Ed. Bulletin 135, Geomembrane Sealing Systems for Dams: Design Principles and Review of Experience. Paris; 2010. p. 464

[3] Giroud JP, Soderman KL. Comparison of geomembranes subjected to differential settlement. Geosynthetics International. 1995;**2**(6):953-969

[4] Giroud JP. Quantification of geosynthetic behavior. Geosynthetics International. 2005;**12**(1):2-27

[5] Giroud JP, Pelte T, Bathurst RJ. Uplift of geomembranes by wind. Geosynthetics International. 1995;**2**(6):897-952

[6] Zornberg JG, Giroud JP. Uplift of geomembranes by wind: Extension of equations. Geosynthetics International. 1997;**4**(2):187-207

[7] Giroud JP, Gleason MH, Zornberg JG. Design of geomembrane anchorage against wind action. Geosynthetics International. 1999;**6**(6):481-507

Chapter 3

Managing the Quality of the Impounded Water

Jonathan A. French

Abstract

Design, construction, and operation of a dam should involve planning and careful consideration not only of the foundation and mass of the dam itself but also of the proper management of the reservoir, and of communities displaced by the reservoir, and impacted in any way upstream or downstream. Many management problems involve a reservoir's density stratification, resulting in low oxygen, phosphorus release, and hydrogen sulfide (H_2S) in the lower layers. Control measures include selective withdrawal and artificial aeration. Case examples are given. Other problems introduced by damming are often best dealt with by measures slow and well-considered, as illustrated by examples. References for further study are provided.

Keywords: stratification, aeration, social impacts

1. Introduction

A dam is the parent of an impoundment, a reservoir, a lake not formerly present. If such a lake develops quality problems, its parent, the dam, bears some responsibility. A range of lake problems can be foreseen, as discussed in the following sections.

Many problems involve the density stratification of a lake (Section 2). Density stratification often leads to thermal stratification, depleted oxygen in the lower layer, and the release from bottom sediments of nutrients causing algal blooms. To address these problems, mitigating measures can be provided in the design and operation of the dam, as well as in control of contaminant inflows to the reservoir (Section 3).

In turn, the mitigating measures usually come with side effects, good or bad, and sometimes costs. Management measures upstream, in, and downstream of reservoirs should be undertaken slowly and carefully (Section 4).

This brief chapter may serve as a literature survey of the management and environmental impact issues introduced.

2. Stratified reservoirs

Many reservoirs, natural or created by dams, become density-stratified.

 IntechOpen

2.1 Definitions

Figure 1 is a schematic profile of a density-stratified reservoir. The dam is on the left. It has a gated spillway in its upper parts, and a valve and metered *withdrawal* exit port at its lowest point, i.e. a low-level sluice. An intake tower with ports at several elevations to *abstract* water for treatment and production is shown at the right in this schematic diagram, although it may actually be placed next to the dam. (Note that usually only one abstraction port is open at a time).

The upper part of the reservoir is the *epilimnion*. Below it lies the slightly denser *hypolimnion*. They are separated by the *pycnocline*, a layer where the rate of change in density with respect to depth is greatest.

Interestingly, the pycnocline is a region that often sees a high concentration of plankton. Harder [1] has observed that plankton, originally homogeneously distributed throughout a volume containing two bodies of water of different densities, will tend to congregate at the density interface, i.e. the pycnocline. Fish are there, too, to feed on the plankton.

2.2 Causes

Density stratification is often a natural phenomenon, due to:

- Solar warming of the epilimnion, rendering it less dense (except at near-freezing temperatures);
- Tributary inflows of cooler water from upstream, which upon entering the reservoir plunge to join the hypolimnion;
- Salinity intrusion from downstream, if the dam is separating the reservoir from the sea or a saline estuary.

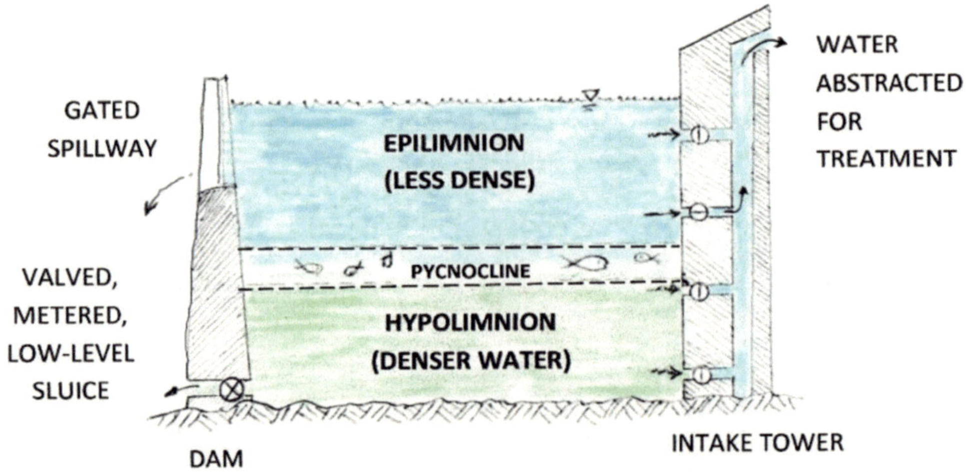

Figure 1.
Schematic profile of a density-stratified reservoir.

2.3 Problems caused by density stratification

In turn, density stratification inhibits vertical mixing, especially across the pycnocline. Impaired vertical mixing leads to the reduction of dissolved oxygen in the hypolimnion. The (impaired) downward diffusion of oxygen through the pycnocline may be estimated as follows: assume the diffusive flux to be $F = -Kdc/dy$, where K is the vertical diffusion coefficient and dc/dy is the vertical gradient of dissolved oxygen concentration. While in general K is a function of vertical gradients in ambient velocity as well as density gradients, Koh and Fan [2] showed that, as a rough approximation in the ocean, K can be related to the vertical density gradient alone:

$$K \approx \left(10^{-4}\right)/e \tag{1}$$

where $e = (1/\rho)(d\rho/dy)$ is the vertical relative density gradient expressed in m^{-1}, and K is expressed in cm^2/s; and where $4 \times 10^{-7}\ m^{-1} \leq e \leq 10^{-2}\ m^{-1}$.

Low DO leads, in turn, to problems such as the release of phosphorus from sediments, which in turn enables algae blooms on the surface. In extreme cases, the reduction in dissolved oxygen results in fish kills, unpleasant odors, and even downright dangerous hydrogen sulfide (H_2S) concentrations onshore near the reservoir, and unattractive quality if hypolimnion water is withdrawn for production. This issue is discussed further by Thomann and Mueller [3], among others.

3. Management measures

3.1 Selective withdrawal, selective abstraction

A reservoir should be able to spill downstream, and not just from a surface spillway. Water for use should be withdrawn, and not just from the reservoir bottom.

If a reservoir is stratified, the poorer quality water is usually (but not always) found in the lower layers.

Figure 2(a) illustrates a discharge strong enough to overcome buoyancy forces both in the hypolimnion and in the pycnocline. The flow pattern approaches that of potential flow in a homogeneous medium. Water from below and above the outlet elevation, perhaps even epilimnion water, is withdrawn. This arrangement is partially satisfactory in the example of Reservoirs C and E, described below, and would greatly improve the product of Reservoir D.

Figure 2(b) illustrates a condition in which withdrawal is exclusively from the hypolimnion, without greatly disturbing the pycnocline. The withdrawal through an outlet port is kept sufficiently small that it "sips" water from a thin layer at the outlet level, somewhat like extracting very few paper sheets from the middle of a stack of papers. The water removed is replaced by the waters above, from the surface down through the epilimnion and the pycnocline to the withdrawal level, descending nearly as a block (except as disturbed by wind or inflows).

Selective withdrawal is discussed in more detail by Imberger in Chapter 6 of Fischer et al. [4], and by Bryant and Wood [5], among others.

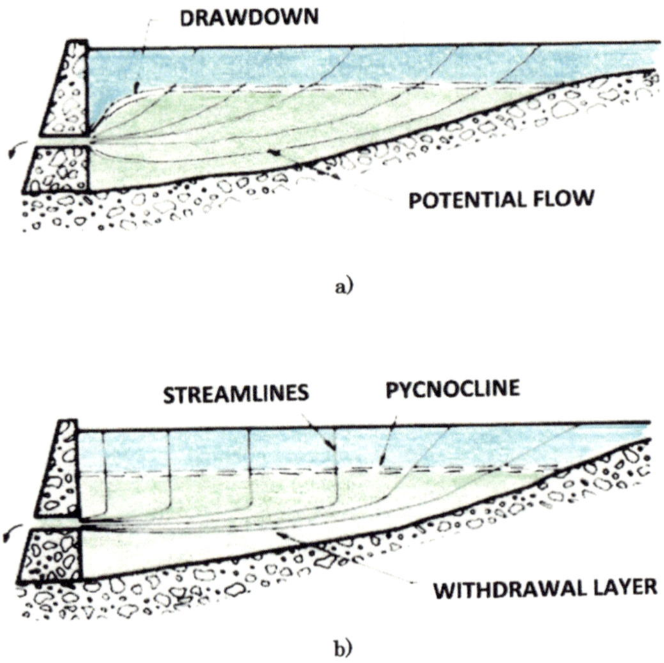

Figure 2.
Withdrawal flow patterns. From Imberger, in Fischer et al. [4]. (a) Discharge is strong enough to overcome buoyancy forces both in the hypolimnion and in the pycnocline. (b) Strong stratification in the hypolimnion, with small discharges.

3.2 Examples

Reservoir A. This artificial body of water, 85 ha in area and 22 m deep, tends to stratify due to solar heating of its epilimnion. Its withdrawal system consists of a tower such as shown at the right side of **Figure 1**, with a choice of several depths from which water can be withdrawn for treatment and production. Epilimnion water is usually of the best quality, but it may be advantageous to withdraw from lower layers, as long as they are acceptable, to keep the hypolimnion from becoming a stagnant body of water with no exit.

As will be discussed below, this reservoir also has an aeration system working in tandem with the withdrawal tower, to keep the lower layers sufficiently aerobic [6].

Reservoir B was formed by damming a tidal estuary. It soon encountered water quality problems, including high salinity in the water withdrawn for production. The problems were attributed to (a) the sill of the gated spillway being set so low, and the gate seals insufficiently tight, that they admitted seawater at high tide; and (b) there was no low-level sluice available to preferentially void the most saline water at low tide. Recognizing these faults led to improved design in a later attempt to convert a tidal estuary to a freshwater reservoir (Reservoir E, below) [6].

Reservoir C receives only freshwater inflows, but thermal stratification leads to low dissolved oxygen in the hypolimnion. Bottom withdrawal for production serves to drain the least good water and keeps the hypolimnion from becoming any worse than it already is. However, the depressed oxygen level is insufficient to prevent the reduction of PO_4, hence, the release of P from sediments. This results in major algal growths on the surface.

Expensive and somewhat problematic large-scale dosing of $CuSO_4$ has been undertaken to control the algal growths. Also, major efforts are undertaken to control the inflow to the reservoir of raw sewage and its oxygen-demanding constituents by upgrading sewerage and treatment in bordering slum areas [7, 8].

Reservoir D, a few kilometers distant from Reservoir C, has a withdrawal point only at mid-depth, near its dam. Its hypolimnion waters are of very poor quality, and lack of a low-level sluice has meant that the hypoxic lower layer could not be easily removed, with taste and odor problems propagating upward even to the water take-off level [6].

Reservoir E: Conversion from seawater to freshwater. A freshwater impoundment can be created by placing a dam across the mouth of a saline tidal estuary to convert it to a freshwater reservoir. When completed, the pool level should be kept higher than the highest local sea tide. The spillway crest is set well above mean tide level, and the spillway gates are made as water-tight as feasible. A low-level sluice is provided to release water from the lowest point in the impoundment out to sea. After the conversion is complete, freshwater for production is to be withdrawn from either of two depths.

The low-level sluice is equipped with a valve, a venturi flow gauge, and a salinometer. Whenever the salinity reading exceeds that suitable for potable water, and there is a net pressure gradient in the sluice towards the sea, the valve can be carefully opened to release the impounded saline water to sea.

This arrangement permits a new impoundment to be converted from seawater to potable freshwater within a time period depending on the rate of freshwater inflow and the volume of the reservoir, without the need to first pump the impoundment dry of its original seawater. At Lower Seletar Reservoir in Singapore, the conversion took about 1 year to accomplish (**Figure 3**) [9].

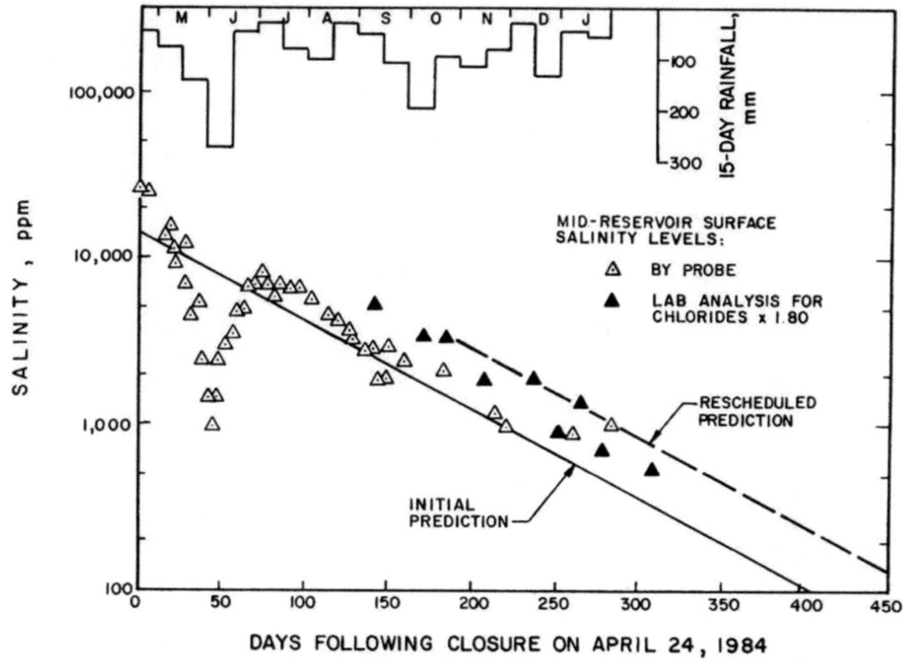

Figure 3.
Measured and predicted rate of desalination in Reservoir E.

3.3 Air-bubble mixing and hypolimnetic oxygenation

A mechanism not directly connected to the dam but often operated in association with withdrawal strategies is air-bubble induced circulation. A relatively simple system consisting of air blowers/compressors on shore, an air-feed tube leading from the blowers to a local deep spot on the lake bed, and an air diffuser at that spot on the lake bed, provide a means both to enhance mixing of epilimnion and hypolimnion waters through the pycnocline and to oxygenate the hypolimnion (**Figure 4**).

In 1970 Cederwall and Ditmars [10] undertook an early study of the mixing of a water body by an air-bubble plume. In 2012 Lima Neto [11] modeled the liquid volume flux in bubbly jets using a simple integral approach. An extensive discussion of air-bubble columns, hypolimnetic oxygenation, and the air flow rates in many actual installations is provided by Cooke et al. [12].

3.4 Examples

Charles River Basin. Originally the Charles River flowing between Boston and Cambridge, Massachusetts, USA had been a saline tidal estuary to Boston Harbor and the Atlantic Ocean, with a 2-m tide range. As a cosmetic measure, a dam was constructed near its mouth ca. 1910 to maintain a consistently high pool level. However, the dam still allowed a very saline hypolimnion to persist—devoid of dissolved oxygen, and rich in hydrogen sulfide. The density difference was so great that boils from the hypolimnion rarely surfaced, and nuisance sulfide odors at the surface were very rare.

However, plans to build a new and better dam led to fears that the density difference would be weakened, allowing more frequent occasions of hypolimnion water reaching

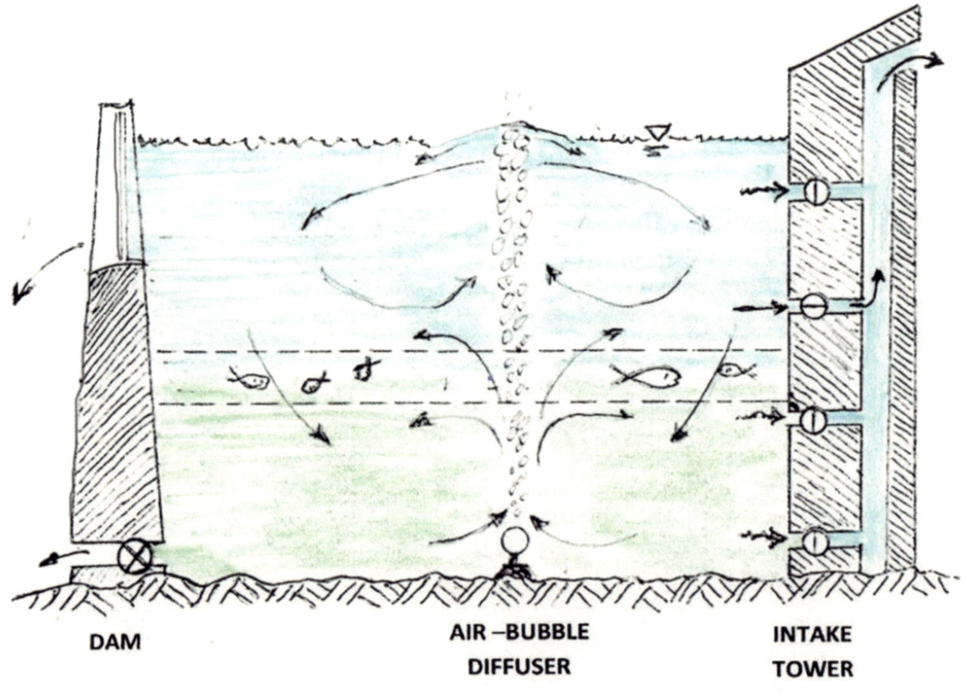

Figure 4.
Air bubble column from air diffuser on lake bed, inducing water circulation patterns.

the surface and causing widespread nuisance and perhaps even dangerous H_2S conditions in the heart of the city. A means was sought to eliminate the H_2S from the lower layer.

Following a successful test, a set of six air-bubble diffusers was deployed over a 3-km length of the Basin. A bed of crushed stone supports each diffuser assembly consisting of a fiberglass cradle and an adjustable rack. The rack holds the diffuser in place and is so constructed that it can be adjusted vertically by about 1 m to provide equal air distribution along the entire length of the diffuser. The diffuser is the 5-m long end of the 75-mm diameter polyethylene pipe from shore. It has four rows of 3-mm diameter holes placed at quarter points around the diffuser pipe's circumference. The holes are spaced 50 mm in the center, and each row is aligned off-center to its adjacent row. A reverse flow check valve is at the intake end of the diffuser and a cleanout flange is provided at the other, normally closed, end. The airflow rate to each of the diffusers is of the order of 0.14 standard m^3/s.

This diffuser system did not fully mix the very strongly stratified upper and lower layers, but it did largely eliminate the H_2S and replace it with some dissolved oxygen [6, 13–15].

Reservoir A. In addition to the multi-level abstraction tower described for Reservoir A above, it was deemed wise to provide a means to ensure vertical mixing for adequate dissolved oxygen at all levels in this 22-m deep impoundment. A single air diffuser, of design similar to those in the Charles River Basin described above, was installed at the deepest point. It successfully kept DO levels adequately high throughout the whole depth and 85-hectare lateral extent of the water body.

Based on this success, similar aerators were installed in other reservoirs in the neighborhood, and have been operating there for nearly four decades [6].

Running costs, it should be noted, especially for electricity, can become significant for an aeration system kept operating on a long-term basis. For most efficient operation, it is best to choose blowers/compressors that are rated for the pressure and flow rate they face. This can be difficult for water depths of the order of 20 m, which make for pressures high for most blowers, but low for most compressors. Again, see the extensive discussion of air-bubble columns, hypolimnetic oxygenation, and the airflow rates in many actual installations by Cooke et al. [12].

An example of aeration failure was the attempt to eliminate H_2S odors emanating from the Kai Tak Nullah, a heavily polluted waterway right next to the runway at Hong Kong's old airport. The failure of a pilot aeration program in the waterway to raise the near-zero dissolved oxygen level was tentatively attributed to the relatively high tidal current to and fro above the diffuser [6].

4. Impacts upstream and downstream of a dam

Large dam projects are often planned and debated for a number of years before being built, in order to optimize their benefits and to minimize their adverse impacts; yet some less desirable results still ensue. Obeng [16] describes four large dam projects in Africa. Lake Kariba was formed by damming the Zambezi River in 1958. Although hydroelectric power was realized and fish grew abundantly in the reservoir, there were problems including population resettlement, deoxygenation, seismic tremors, an explosive bloom of *Salvinia auriculata*, and schistosomiasis. The impacts on local populations by the Kariba Dam on the Zambezi River bordering Zambia and Zimbabwe have been studied extensively over the past half-century by Dr. Thayer Scudder and his associates [17].

Kainji Dam on the Niger River in Nigeria was intended for hydroelectric power, flood control, navigation, irrigation, and fisheries [16]. It has met many of these objectives; advance planning has helped to mitigate, although not entirely obviate adverse effects such as algal blooms and schistosomiasis.

Lake Nasser, formed by The Aswan High Dam on the Nile River in Upper Egypt formed an impoundment known as Lake Nasser in Egypt and Lake Nubia in Sudan. As intended, the dam has contained floods, produced hydroelectricity, stored water, and contributed to fisheries, but it is the adverse effects: environmental, affecting the health of the people, and threatening the limestone foundations of many of the Pharaonic monuments that have brought the most criticism of the project [6, 16].

The dam project on the Volta River in Ghana, proposed as far back as the early 1920s, was well studied and planned. However, perhaps due to pressing economic interests, the forestalling of possible environmental impacts did not receive adequate attention during construction [16]. The resettlement of some 80,000 displaced people, though well planned beforehand, was in the event rushed and poorly executed. Eventually, however, the dam fulfilled the intentions for which it was built.

For another example, in 2021 the quick filling of the reservoir behind the dam on the Blue Nile in Ethiopia has been of heated international concern to Sudan and Egypt [18]. Öziş [19] has discussed the quantity and quality implications in the upstream and downstream dimensions of damming on the Euphrates and Tigris Rivers in the Middle East. Anderson [20] discusses the impacts of the Three Gorges Dam on the Yangtze River in China.

The operators of a reservoir should anticipate a large storm inflow and may decide to draw down the reservoir to accommodate it. However, the suddenness of the drawdown can leave riparian groundwater levels unbalanced by the reservoir, to the extent that major landslides into the reservoir cannot be held back, and the suddenly displaced water overtops the dam and causes catastrophic flooding downstream. A prime example is the Vajont Dam flood near Belluno, Italy in 1963, caused by 260 million cubic meters of rock sliding abruptly into the reservoir [21].

A low coastal dam can be called a barrage. Burt & Watts [22] have edited the proceedings of an international conference held in 1996 in Cardiff, UK titled, "Barrages: Engineering Design and Environmental Impacts." The 39 professional papers cover Concepts and Issues; Environmental Implications; Changing Estuaries; Cardiff Bay's Hydraulic Regime; the Cardiff Bay Barrage; Ecological Issues; Water Quality; Costs, Benefits and Decisions; Tawe Barrage; River Flows and Control; and Planning and Control.

5. Conclusion

Density stratification often happens normally in a natural lake or an impoundment behind a dam, due to factors such as solar heating of the upper layers or colder or saltier water intruding the lower layers. Density stratification inhibits vertical turbulent mixing, particularly across the pycnocline, the elevation where the change in density with respect to depth is greatest.

Inhibited vertical mixing can lead to depressed dissolved oxygen levels in the hypolimnion (lower layers), particularly if the lake is subject to sewage or other oxygen-demanding inflow. Low oxygen, in turn, releases phosphorus from sediments to feed algal and other phyto-blooms on the surface.

The adverse effects of stratification can be ameliorated by the selective withdrawal of water, to waste, or to abstraction, from the upper or the lower layers, as deemed wise. Air-bubble columns have been used to reduce the intensity of stratification and to increase the oxygen level in the lower layers.

It is important to try to anticipate the possible adverse impacts that the construction of a dam, and the creation of its impoundment, may have on the affected population, environment, and geology nearby, upstream, and downstream. As discussed above, the possible impacts are many and varied. Field conditions from site to site may vary sufficiently that simple import of one solution to a problem, successful at one place, may not necessarily work well at another, as noted for Kai Tak Nullah. The best advice is to move forward slowly and carefully with implementation.

Author details

Jonathan A. French
Retired; Formerly with CDM Smith, Inc., Boston, MA, USA

*Address all correspondence to: jonafrench@aol.com

References

[1] Harder W. Reactions of plankton organisms to water stratification. Limnology and Oceanography. 1968; **13**(1):156-168

[2] Koh RCY, Fan L-N. Mathematical Models for the Prediction of Temperature Distributions Resulting from the Discharge of Heated Water into Large Bodies of Water. USEPA Water Quality Control Office Contract No. 14-12-570. Washington, D.C., USA: U.S. Government Printing Office; 1970

[3] Thomann RV, Mueller J. Principles of Surface Water Quality Modeling and Control. New York, NY, USA: Harper and Row; 1987

[4] Fischer HB et al. Mixing in Inland and Coastal Waters. New York, NY, USA: Academic Press; 1979

[5] Bryant PJ, Wood IR. Selective withdrawal from a layered fluid. Journal of Fluid Mechanics. 1976;**77**(3):581

[6] French JA. Personal Communication

[7] French JA et al. Remediation for Guarapiranga Reservoir, Sao Paulo, Brazil. In: Fifth Stockholm Water Symposium. Abstracts. Stockholm, Sweden. 1995. p. 75

[8] Oliveira LHW et al. Dimensionamento de Sistema de Aeração por Circulação Artificial da Coluna d'Água para o Reservatorio Guarapiranga, em São Paulo. 19th Congresso Brasileiro de Engenharia Sanitária e Ambiental. Brasil: Foz do Iguaçu; 1996. In Portuguese

[9] French JA, Harley BM, Neysadurai A. Desalination of an impounded estuary. In: Proc. of the 1985 Specialty Conference in Environmental Engineering. Boston, Massachusetts: ASCE; 1985

[10] Cederwall K, Ditmars JD. Analysis of air-bubble plumes. Rep. No. KH-R-24. W.M. Keck Laboratory of Hydraulics and Water Resources, Division of Engineering and Applied Physics, California Institute of Technology; 1970

[11] Neto L, Iran E. Modeling the liquid volume flux in bubbly jets using a simple integral approach. Journal of Hydraulic Engineering ASCE. 2012;**138**(2):210

[12] Cooke GD et al. Restoration and Management of Lakes and Reservoirs. 3rd ed. Baton Rouge, FL, USA: CRC/Taylor and Francis; 2005

[13] Metropolitan District Commission. Charles River Artificial Destratification Project, Boston, Massachusetts. 1981

[14] Science Museum of Boston. Handout: Why are Those Bubbles in the River? 1981

[15] Godbey A et al. An impact assessment of the new Charles River dam. Boston's Charles River Basin: An Engineering Landmark, Journal of the Boston Society of Civil Engineers Section, ASCE. 1981;**67**(4):362-385

[16] Obeng L. Environmental impacts of four African impoundments. In: Gunnerson C, Kalbermatten J, editors. Environmental Impacts of International Civil Engineering Projects and Practices. ASCE Preprint 2920. Reston, VA, USA. 1977. pp. 29-43

[17] Matanzima J. Thayer Scudder's four stage framework, water resources dispossession and appropriation: The Kariba case. International Journal of

Water Resources Development. 2021;**38**(2):322

[18] Saied M. Ethiopia preps for third filling as Nile dam diplomacy fails. Al-Monitor. October 20, 2021

[19] Öziş Ü. Quantity/quality implications in the upstream/downstream dimensions of Euphrates & Tigris in the Middle-East. In: Fifth Stockholm Symposium. Stockholm, Sweden. 1995

[20] Anderson W (Schoolworkhelper Editorial Team). Impacts of Three Gorges Dam. Poster in: SchoolWorkHelper. 2019. Available from: https://schoolworkhelper.net/impacts-of-three-gorges-dam/

[21] Graf von Hardenberg W. Expecting Disaster: The 1963 Landslide of the Vajont Dam. Arcadia Collection of Technology and Expertise. No. 8. 2011

[22] Burt N, Watts J. Barrages: Engineering Design & Environmental Impacts. Chichester, UK: Wiley; 1996

Chapter 4

Analysing the Possibility of Failure of Cascade Dam System and a Case Study from Brazil

Angélica Luciana Barros de Campos,
Ruben Jose Ramos Cardia and Welitom Ttatom Pereira da Silva

Abstract

A cascade dam system poses more hazards for downstream life and structures, when compared with a single dam located on a river. Therefore, there is a need to develop differentiated procedures to classify and regulate these dams. In the state of Mato Grosso (MT), Brazil, it is common to find multipurpose dams, which can be considered as a cascade, when a dam failure causes adverse effects in downstream dams. The objective of the study is to analyse the possibility of dam failure located in the cascade system operated by the municipality of Várzea Grande, MT by the Associated Potential Damage (APD) classification used throughout the country. In order to do this, the specification namely "Simplified Methodology to Define the Classification Flood Zone of Associated Potential Damage of a Dam" developed by the National Laboratory for Civil Engineering in Portugal (LNEC in Portuguese) was utilised. This specification was adapted by the National Water and Sanitation Agency (ANA in Portuguese) in Brazil. In the case study, there are three dams (Dam 1, Dam 2 and Dam 3) in the cascade system. Dam 1 can cause overtopping problem for Dam 2 and Dam 3. According to APD classification, dams considered for the study are categorised as "high dam".

Keywords: dam safety, dam failure, dam classification

1. Introduction

The increase in the number of dams has increased the concern with the safety of these structures [1]. Major accidents with dams have drawn attention to this problem, as failures in these structures can cause irreversible environmental damage, as well as human and animal deaths.

Advantages in the implementation of cascade dam projects have made this conception commonplace and risk analysis a complex activity. The advantages of these projects include simplifications derived from decentralisation (lower cost, implementation time and complexity to comply with legislation); the possibility of storing water to generate electricity in downstream dams; the transformation of stretches that are not navigable into waterways; full use of the flow of water; the development of rolling projects; and coordinated dispatch of energy [2–4].

Within this context, numerous studies have been developed to provide subsidies for operation, maintenance and decision-making in the area of dams. Some of these studies are briefly presented in the sequence. In a work on the Dadu River basin, one of the main rivers in China, transmission risk and overtopping between dams in cascade reservoirs were studied [3]. Dam risk was divided into two parts: own risk and additional risk. Based on this, a methodology that considers the relevant concepts and equations for calculating the risk probability of dams in cascade reservoirs was proposed. The results indicated that the proposed methodology realises an effective connection with the risk calculation of the single-reservoir dam method, which can provide a reference for the risk assessment and management of cascaded reservoirs in the basin [3].

The research focused on the genetic-algorithm-based optimisation of rule curves to maximise hydropower and associated benefits through cascading reservoirs [4]. The optimisation was performed using RESOOSE model and applied to multipurpose Diamer Basha and Terbela reservoirs on Indus River in Pakistan. According to [4], the results show that hydropower was optimised along with the maximisation of associated benefits (power generation, irrigation and sediment discharge).

A study to describe the principles of dam risk classification methods, the International Commission on Large Dams (ICOLD) method and the Bureau method, and comparatively discussing total risk classifications was carried out by Topcu and Tosun [5]. In conclusion, the following stand out: the safety assessment for cascade dams should be different from that of a single dam; the rupture of any dam in a cascade structure may result in overtopping or breakage of the dam or dams downstream; the probability of rupture of any dam in a cascade structure is higher than that of a single dam; and a new safety concept should be developed for dams in a cascade system [5].

2. Theoretical background

Some topics considered important for the study of dam safety in Brazil are presented in the following.

2.1 National dam safety policy

In this context, the National Dam Safety Policy (PNSB in Portuguese), established by [6] and amended by [7] aims to ensure that dam safety standards are followed (mainly by developers) to reduce the possibility of accidents and their consequences, as well as regulate safety actions and standards. The policy defines a dam as any structure built inside or outside a permanent or temporary watercourse, in a thalweg or in a diked pit, for the purpose of containment or accumulation of liquid substances or mixtures of liquids and solids, comprising the dam and associated structures.

2.2 Dam safety plan

An important instrument of the National Dam Safety Policy, which must be implemented by developers, is the Dam Safety Plan (PSB in Portuguese), whose objectives are to assist and serve as a planning tool in the dam safety management. In the Dam Safety Plan structure, the information and technical data of the dam referring to construction, operation, maintenance and the overview of the current state of safety, obtained through regular periodic inspections, must be documented. For multipurpose dams inspected by the National Water and Sanitation Agency, this agency issued

[8] establishing the implementation of procedures, including the minimum content and level of detail of the Dam Safety Plan.

Depending on the conditions of each dam, the Dam Safety Plan may require preparing up to six volumes, distributed as follows: Volume I—General Information; Volume II—Technical Documentation of the Project; Volume III—Plans and Procedures; Volume IV—Records and Controls; Volume V—Periodic Review of Dam Safety (PRDS); Volume VI—Emergency Action Plan (PAE), when required.

2.3 Dam classification

Dams can be built for various uses, including irrigation, aquaculture, human water supply, industrial water supply, animal watering, leisure, fishing, electricity generation and disposal of mining and industrial tailings, among others.

Dam classification consists of an instrument of the National Dam Safety Policy and is a procedure normally carried out by inspection bodies, following the criteria of [9], and was established by the National Water and Sanitation Agency, the Risk Matrix and Associated Potential Damage. It is known that currently (in 2022), there are negotiations to be carried out with the updating and adequacy of the Brazilian National Council of Water Resources (CNRH) definitions and recommendations.

According to the aforementioned law and its amendments, the definition refers to the classification of the dam risk category (DRC) by the associated potential damage (APD) and by its reservoir volume. In the DRC, aspects that may influence the possibility of an accident or disaster are verified, taking into account the technical characteristics and the state of conservation, and the Dam Safety Plan, that is, the conditions of the risk of dam accidents due to failure or lack of adequate actions, is the developer's responsibility.

On the other hand, the APD refers to damage that can occur due to failure, leakage, infiltration into the soil or malfunction of a dam, regardless of its probability of occurrence, to be graded according to the loss of human life and social impacts, economic and environmental.

The dams must be classified into five classes (A, B, C, D and E, presented in (**Table 1**), and those that present a higher degree classification, in the risk category scale and associated potential damage, must prepare a more complete plan.

Currently, Brazil has 22,649 dams registered by 33 inspection bodies in the National Information System on Dam Safety (SNISB in Portuguese). There are 6507 dams assessed by DRC, and of these 2503 were classified as high DRC; there are 1388 dams assessed by APD, and of these 871 were classified as high APD.

2.4 Situation of cascade dams

The state of Mato Grosso, located in the Central-West region of Brazil, has five Dam Safety inspection bodies: (1) Mato Grosso State Secretariat for the Environment

Risk category	Associated potential damage (APD)		
	High	**Medium**	**Low**
High	A	B	C
Medium	A	C	D
Low	A	C	E

Table 1.
Risk matrix and associated potential damage.

(SEMA-MT in Portuguese) inspects dams intended for multiple uses, located in state rivers; (2) National Water and Sanitation Agency (ANA in Portuguese) inspects dams intended for multiple uses, located in federal rivers; (3) National Mining Agency (ANM in Portuguese) inspects dams destined for the disposal of mining tailings; (4) National Electric Energy Agency (ANEEL in Portuguese) inspects dams intended for generating electric energy; (5) Brazilian Institute of Environment and Renewable Natural Resources (IBAMA in Portuguese) inspects dams intended to contain industrial waste. However, only four of these bodies are effectively inspected, except for IBAMA.

According to the National Information System on Dam Safety, there are currently 652 dams registered in the state of Mato Grosso. Of these, 262 are inspected by the Mato Grosso State Secretariat for the Environment, 132 by the National Water and Sanitation Agency, 141 by the National Mining Agency and 117 by the National Electric Energy Agency.

Dam safety has been a very important item for a long time, although in Brazil, the development of regulatory actions only took place after Law No. 12,334, on November 20, 2010. The improvement of actions evolved after the Inspection Agencies (federal and state) issued the respective regulations.

After establishing the National Dam Safety Policy, there were advances in dam classification and regularisation in Brazil. However, progress is still needed in terms of inspectors' performance, specifically in dam regularisation (issuance of grants and environmental licences, among others), construction and refinement of registries, inserting dams into the National Information System on Dam Safety, dam classification, reviewing regulations [6] and on-site and documentary inspection actions. Most dams were not regularised and/or included in the National Dam Safety Policy by the inspection entities due to a lack of basic information such as height and volume, or because they were not classified as to the APD, or because of the difficulty in identifying the responsible developer [10].

As in several states in Brazil, the state of Mato Grosso also has problems regarding the safety of multipurpose dams located in state water bodies, as presented in [11]. This is because the competence in the state for managing civil works, such as dam construction, is carried out by another sector of the Mato Grosso State Secretariat for the Environment, which deals with environmental licencing, requiring strong coordination between the Water Resource Superintendence and other areas of the secretariat to carry out the actions proposed by the National Water and Sanitation Agency, as in the case of dam regularisation in terms of licencing, granting and dam classification, together with the fact that the state also has recently published regulations for dam regularisation.

Most multipurpose dams, located in existing state water bodies in the state of Mato Grosso, were predominantly built on land; they have a small reservoir area and height and are mostly characterised as small dams. They are mostly located on rural properties, with predominant uses for irrigation and aquaculture activities. These dams are, in many cases, maintained (not so adequately) by small developers who usually do not have sufficient technical knowledge for more accurate assessments and/or financial resources to hire qualified professionals. This situation is common in several regions of Brazil and in other parts of the world [12].

It was noted that the vast majority of them are located mainly on farms, but far from the sites where the caretakers are. Thus, there is little or no concern with dam safety on the part of developers, mainly due to the absence of specific regulation, until mid-2017, and, consequently, a lack of guidance, a lack of disclosure and notification and/or lack of knowledge. Most dams do not have a technician responsible for the structure and were built without having a project, without constructive care and without environmental licencing. Some were built for access purposes, on irregular terrain and ended up damming (rain) water. These normally do not present dam aspects, but they present risks because of the lack of adequate executive and maintenance conditions.

The dams in the best safety conditions are those that have the "regular road function". In this case, there is no excess vegetation, at least at the crest of the dam. It can also be observed that the excess vegetation, a common characteristic found in the dams of the state, is part of a culture of the developers and employees of the farms, who must believe that the excessive vegetation (mainly trees on the crest) on the slopes would protect the dam. However, it is the opposite.

In the visits to the dams, it was noticed that some conditions are recurrent in the state's dams and require care and maintenance, such as excess of high vegetation on the crest and slopes; crest cracks; fissures and erosion on the slope; infiltration moisture, on the downstream slope; and structures close to roads, which can cause socio-environmental impacts (and even deaths) in the event of a possible failure.

Most of the dam accidents, that had a loss of life, occurred in small and/or medium dams, with a height of less than 30 meters. Therefore, these dams have become a matter of concern and of greater interest to dam safety professionals [13].

In the state of Mato Grosso (MT), it is common to find these small dams located in the same watercourse (or in thalwegs), which can be sequential (when there is no negative effect of a dam on the downstream dam) or in cascade. Many are located close to highways and/or other infrastructure, which may represent high APD.

Dam failure is considered as a cascade, when the occurrence of an extreme event (with loss of resistance, or overtopping and/or overflowing) of one of them, generating a burst flood wave, may indicate that it serves as a cause for the possibility of failure of another dam, existing downstream (in the same watercourse or thalweg).

Each developer should manage the dams they own properly, so as not to assign additional risks to other dams in the cascade, which may even belong to other developers. The flood wave caused by a failure is generally greater than the design flow and capacity of the dam spillway or spillway immediately downstream, as the reservoir volume is released in a short time interval. This flood wave can cause the downstream dam to overflow ("overtopping") and failure [14].

Given that in the state of Mato Grosso, there have already been cascading failures, which is a concern for the inspection body, there is a need to develop differentiated procedures for the regularisation and classification of these structures. Thus, the objective of the study is to analyse the possibility of failure, according to the conditions of dams located in the same watercourse (or thalweg), in the municipality of Várzea Grande, MT. APD classification is also made possible of these dams and, thus, presents a procedure (or protocol) for the APD classification of dams located in the same body of water (or thalweg).

3. Methodology

To analyse the negative influence of a dam failure in other dams located in the same watercourse, the "Simplified Methodology to Define the Classification Flood Zone of Associated Potential Damage of a Dam" was used. This methodology is currently used by the National Water and Sanitation Agency to classify dams within the competence. It is a methodology developed at the National Laboratory for Civil Engineering in Portugal for the National Water and Sanitation Agency, which generates a reference area or polygon for APD estimation. The methodology assumes that the available data from both the dams and the relief do not allow the accurate modelling of the failure. Thus, the appropriate methodology was simplified so as to use the Shuttle Radar Topography Mission (SRTM) as a digital elevation model and a minimum of input data without describing the geometry of both the dam and the downstream flow channel.

Therefore, the objective of this area is only to delimit the area of analysis to classify the APD. It is similar to a flooded area; however, we cannot understand it as an exact flood area, due to insufficient data precision, especially the relief data (30 m SRTM). It is limited to being a reference on which the public agent, or the dam safety supervisory body, can be based to prioritise its performance [15].

According to [15], the methodology (considering the area downstream of the dam and some cross sections, reference and control) consists of the following procedures:

- Empirical calculation of the length of the river to be modelled.
- Verification of the adequacy of the determined limit, analysing the occupation and geomorphology of the valley downstream of the dam.
- Calculation of the maximum flow associated with the dam failure.
- Calculation of peak flows in each of the cross sections along the valley.
- Comparison of the damped flow in each cross section along the valley with the maximum flow of the spillway.
- Obtaining the altimetry of points along the cross sections.
- Hydraulic calculation to estimate the maximum level of the failure wave in each of the cross sections.
- Creation of the surface surrounding the failure wave.
- Obtaining the polygon, for the APD classification.
- Possible adoption of adjustments, to consider specific characteristics of the site.

To generate the area, it is necessary to use the ArcGis software. In the study, version 10.5 was used. In addition, the following additional data are required:

- SRTM 30 m.
- Reservoir volume (hm^3).
- Height of the dam (m).
- Crowning quota (meters above mean sea level, m).
- Satellite images.

The risk analysis of cascading failure of the dams (selected) occurred from thematic maps with the classification areas, where they were evaluated according to the aspects of the classification of the APD presented in [16], such as total volume of the reservoir; potential for loss of human life; environmental impact and socio-economic impact.

For the analysis, a case study was carried out using dams 1 (B1), dam 2 (B2) and dam 3 (B3) located in Ribeirão Esmeril, a municipality of Várzea Grande, Mato Grosso and presented in **Figure 1**.

The structures were built using homogeneous land as material, with priority use for aquaculture activity. These dams are registered with the Mato Grosso State Secretariat for the Environment Water Resource Superintendence. They were inspected in 2016 and have data collected in the field by technicians from the inspection body (SEMA-MT).

The location and height of the dams were obtained from the Mato Grosso State Secretariat for the Environment survey report. The volume was estimated from the wetland and height of the dam. Additional information on the dams includes: foundation on soil; embankment dam formed of compacted earth, material homogeneous; construction date, unknown; volume of dam 1 = 0.24 hm^3; volume of dam 2 = 0.50 hm^3; volume of dam 3 = 0.39 hm^3; height of dam 1 = 2.7 m; height of dam 2 = 4.8 m; and height of dam 3 = 6.3 m. The crest of the dams was obtained from Google Earth. The input data used in the simulations are shown in **Table 2**.

In the simulations, to analyse the negative influence of dams located in the same watercourse, some scenarios were considered, where they could occur:

Figure 1.
Dams used to analyse the negative influence of dams located in the same watercourse.

Simulation	Scenario	Estimated volume (hm^3)	H (m)	Crowning quota (m)
S1	Isolated failure of B1 dam	0.24	2.7	175
S2	Isolated failure of B2 dam	0.50	4.8	174
S3	Isolated failure of B3 dam	0.39	6.3	173
S4	Failure of B2 and B3 dams	0.89	6.3	173
S5	Failure of B1, B2 and B3 dams	1.13	6.3	173

Table 2.
Input data for generating the APD classification area, considering different scenarios.

- Isolated failure of each of the dams, according to the recommendations of [15].

- Cascade failure of dams B2 and B3, where the initial failure occurs in dam B2, but given the limitations of the application, it is considered as occurring in dam B3, adding the volumes of water stored in the two structures (volume of dam B2 added to the volume of the B3 dam), as if it were only in the B3.

- Failure of the three dams. In this case (given the limitations of the methodology), the modelled failure occurs in dam B3 and the volume considered in the simulation was the sum of the volumes of the three dams.

As the methodology for generating the classification area can only consider data from only one dam, for the cascade failure scenarios (simulation S4 and S5), the simulation was started at the last dam where the failure would occur.

Therefore, the height and crown height data used in the S4 and S5 simulations were the data from the B3 dam, while the volume used was the sum of the three reservoirs.

4. Results and discussion

Figure 2 shows the classification area considering only the B1 failure.

By analysing the isolated areas and the values of the maximum levels obtained, it can be observed that the failure of the upstream dam (dam B1) would probably cause the failure of dams B2 and B3, as the maximum levels obtained in the simulation of dam B1 in the sections of dams B2 and B3 are 175.32 m and 174.33 m, respectively. The heights of the crest of these dams are 174 m and 173 m, respectively.

Thus, by the simulation, the water could pass over the crest of these dams, which would probably cause a sequential failure by overtopping.

A similar situation occurred with the maximum levels of the isolated failure simulation of dam B2. In this case, the wave would probably cause water to pass over dam B3, according to the simulation, as the maximum height of the flood wave in the section where dam B3 is located is 174.89 m and the crest level of that dam is 173 m.

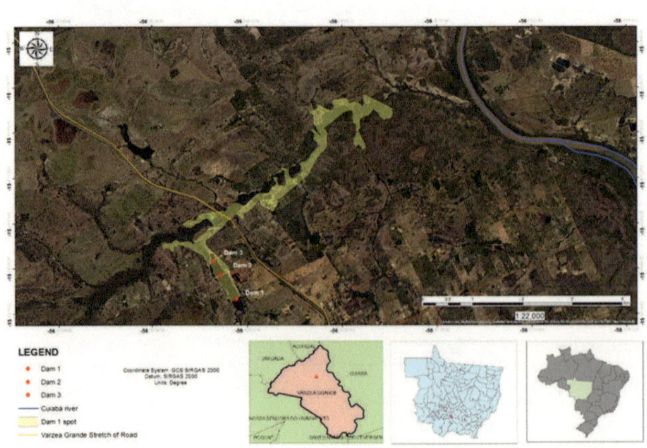

Figure 2.
Area generated considering simulation 1.

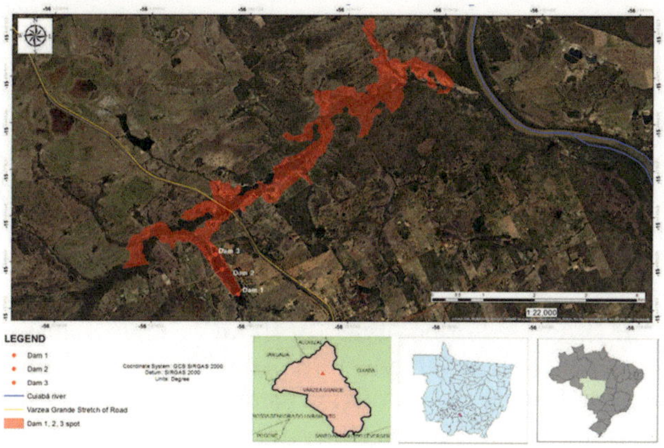

Figure 3.
Area caused by the hypothetical failure of dams B1, B2, and B3.

Figure 3 shows that the areas generated by the cascade failures are larger than the areas considering the isolated dam failures.

This demonstrates the need for a differentiated regularisation procedure for these dams, since they must be classified together, as they generate greater negative consequences, that is, possible loss of life, damage to the environment and socio-economic losses.

In the case studied, according to the simulation of the hypothetical failure area, a possible failure could affect housing and the federal highway located downstream of the dams, as well as fish farming ponds, also generating economic impacts.

Considering the dam classification separately, all are classified with HIGH APD, according to [9]. Following the classification matrix regarding the APD of resolutions [8, 16], the area in the three cases passes through housing where there is permanent

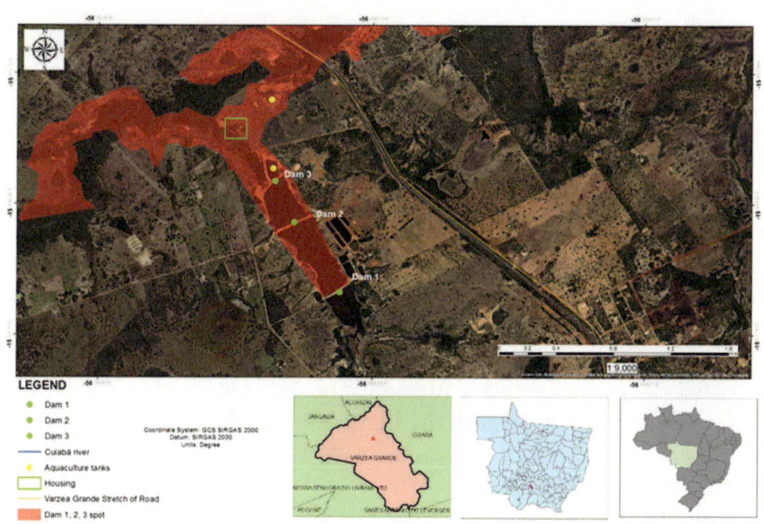

Figure 4.
Detail of the housing, aquaculture tanks and road section present in the area, considering the failure of dams B1, B2, and B3.

occupation, a federal highway and aquaculture tanks, as shown in **Figure 4**, totalling APD equal to 17 (a = 1, b = 12, c = 1, and d = 3, in the evaluation table).

According to [9], there was guidance to already consider HIGH APD, in case any item of the "Conservation State" is greater than or equal to "10".

The same occurred considering the failure of the cascade dams, both for the scenario of the failure of dams B2 and B3 and for the failure scenario of the three dams.

It is important to mention that the program itself calculates the length of the river to be modelled. For all simulations, the limit distance calculated by the program was approximately 7 km.

Therefore, all the generated flood areas are about 7 km long and, consequently, would appear (unconfirmed) that they would probably not reach the Cuiabá river.

However, it is believed that if the simulation were carried out for a greater distance, the area would reach the main body of water. However, it is also believed that the Cuiabá river itself (as it is larger) would be able to lessen this cascade failure flood wave.

From these simulations, a procedure to analyse the APD of cascade dams was devised, which consists of the following steps:

- Carry out the simulation considering the isolated dam failure.

- Check if there would be a situation of overtopping of the downstream dams, caused by the upstream dams, considering the maximum elevation values of the classification area.

- Carry out the simulation considering the extreme case of cascading failure of all dams. In this case, the simulation must use data from the last downstream dam (height and height of the crest), but with the total volume of the three dams.

- Analyse the APD classification criteria presented in [16], based on the APD classification area generated by the simulation that considers the extreme case of a cascade failure.

- Check the APD rating (high, medium or low) as per **Table 1** from [9], showed in item Theoretical background.

5. Conclusion

The existence of numerous small dams and many in cascade spread throughout the state of Mato Grosso is a risk of accidents and catastrophes, related to the safety of dams. From the way they originated and live day to day (lack of design, inadequate construction, lack of knowledge of technique and legislation, lack of maintenance, lack of resources, etc.), there is uncertainty about their safety.

Thus, the analysis of the negative influence of dams located in the same watercourse is extremely important so that the inspection bodies of dam safety are aware of the possible impacts of dams in cascade and not just sequential when there would be no failure in the "domino effect".

The development of related studies and their regulation will make it mandatory for designers to provide and submit to the supervisory body: studies, reports and action plans, necessary to guarantee the safety of these dams, as well as immediate resilience activities, thus guaranteeing the safety of the population and the environment.

Acknowledgements

Both the authors wish to express their thanks to the Institute of Applied Economic Research (IPEA) and National Water and Sanitation Agency (ANA) for the scholarship to researchers, through the Project "Aperfeiçoamento de Ferramentas Estaduais de Gestão de Recursos Hídricos no Âmbito do Progestão". They are also thankful to the State Environment Secretary, Mato Grosso State (SEMA-MT) for the technical support, support for field activities, as well as for providing and authorising the use and dissemination of data. The support of the Federal University of Mato Grosso – "Edital de Apoio à Pesquisa (2021)".

Conflict of interest

The authors declare no conflict of interest.

Author details

Angélica Luciana Barros de Campos[1], Ruben Jose Ramos Cardia[2] and Welitom Ttatom Pereira da Silva[3]*

1 National Agency of Mining, Cuiabá, Brazil

2 RJC Engineering and Fractal Engineering, Bauru, Brazil

3 Federal University of Mato Grosso, Cuiabá, Brasil

*Address all correspondence to: wttatom@gmail.com

References

[1] Fontenelle MC, Fontenelle AS, Matos YMP, Monteiro FF. Risk assessments on dams: a case study of the Malcozinhado dam in north-east Brazil (Avaliações de risco em barragens: estudo de caso da barragem Malcozinhado no nordeste brasileiro). Revista Eletrônica de Engenharia Civil. 2018;**14**:25-42. DOI: 10.5216/reec.v14i1.46356

[2] Amazônia Real. Aluminium and Dams: 23 - Cascade Dams (Alumínio e Barragens: 23 – Cascatas de barragens) [Internet]. 2016. Available from: https://amazoniareal.com.br/aluminio-e-barragens-23-cascatas-de-barragens/ [Accessed: April 01, 2022]

[3] Wang T, Li Z, Ge W, Zhang Y, Jiao Y, Sun H, et al. Calculation of dam risk probability of cascade reservoirs considering risk transmission and superposition. Journal of Hydrology. 2022;**609**:127768. DOI: 10.1016/j.jhydrol.2022.127768

[4] Rashid MU, Abid I, Latif A. Optimization of hydropower and related benefits through Cascade reservoirs for sustainable economic growth. Renewable Energy. 2022;**185**:241-254. DOI: 10.1016/j.renene.2021.12.073

[5] Topcu S, Tosun H. An overview on total risk classifications for dams. In: Proceedings of the 5th Congress for Dams; 30 September-2 October 2005. Struga: Macedonian Committee on Large Dams (MACOLD); 2021. pp. 1-9

[6] BRASIL. Lei n° 12.334, de 20 de setembro de 2010. Establishes the National Policy on Safety of Dams Intended for Water Accumulation for any uses, for Final or Temporary Disposal of Tailings and for Accumulation of Industrial Waste, Creates the National Information System on Dam Safety (Estabelece a Política Nacional de Segurança de Barragens destinadas à acumulação de água para quaisquer usos, à disposição final ou temporária de rejeitos e à acumulação de resíduos industriais, cria o Sistema Nacional de Informações sobre Segurança de Barragens) [Internet]. 2010. Available from: http://www.planalto.gov.br/ccivil_03/_ato2007-2010/2010/lei/l12334.htm [Accessed: January 31, 2022]

[7] BRASIL. Lei n° 14.066, de 30 de setembro de 2020. Revision and Amendment of Law no. 12,334 of 20 September 2010, Which Established the National Policy on Dam Safety (PNSB), Intended for the Accumulation of Water for any Uses, the Final or Temporary Disposal of Waste and the Accumulation of Industrial Waste, Creates the National Information System on Dam Safety (SNISB); Law no. 7. Law No. 797 of 10 July 1989, Which Created the National Environmental Fund (FNMA); Law No. 9.433 of 8 January 1997, Which Established the National Water Resources Policy; and Decree-Law No. 227 of 28 February 1967 (Mining Code) (Revisão e alteração da Lei n. 12.334,de 20 de setembro de 2010, que estabeleceu a Política Nacional de Segurança de Barragens (PNSB), destinadas à acumulação de água para quaisquer usos, à disposição final ou temporária de rejeitos e à acumulação de resíduos industriais, cria o Sistema Nacional de Informações sobre Segurança de Barragens (SNISB); a Lei n° 7.797, de 10 de julho de 1989, que criou o Fundo Nacional do Meio Ambiente (FNMA); a Lei n° 9.433, de 8 de janeiro de 1997, que instituiu a Política Nacional de Recursos Hídricos; e o Decreto-Lei n° 227, de 28 de fevereiro de 1967 (Código de Mineração)) [Internet]. 2020. Available from: http://www.planalto.gov.br/

ccivil_03/_ato2019-2022/2020/lei/L14066.htm [Accessed: January 31, 2022]

[8] ANA – Agência Nacional de Águas e Saneamento Básico. Resolution No. 236 of 30 January 2017 (Resolução n° 236, de 30 de janeiro de 2017) [Internet]. 2010. Available from: https://arquivos.ana.gov.br/resolucoes/2017/236-2017.pdf [Accessed: January 31, 2022]

[9] BRASIL. Resolution No. 143 of 10 July 2012 (Resolução n° 143, de 10 de julho de 2012) [Internet]. 2012. Available from: https://cnrh.mdr.gov.br/resolucoes/1922-resolucao-n-143-de-10-de-julho-de-2012/file [Accessed: January 31, 2022]

[10] ANA – Agência Nacional de Águas e Saneamento Básico. Report on dam safety (Relatório de segurança de barragens) [Internet]. 2016. Available from: https://www.snisb.gov.br/portal/snisb/relatorio-anual-de-seguranca-de-barragem/rsb-2016/relatorio-de-seguranca-de-barragens-2016.pdf/view [Accessed: November 15, 2018]

[11] IPEA – Instituto de Pesquisa Econômica Aplicada. Program to Consolidate the National pact for Water Management (1 cycle) - State of Mato Grosso (Programa de consolidação do pacto nacional pela gestão de águas (1 ciclo) - Estado de Mato Grosso) [Internet]. 2017. Available from: http://www.ipea.gov.br/portal/images/stories/PDFs/relatorio_institucional/171214_relatorio_institucional_fronteiras_do_brasil_mato_grosso.pdf [Accessed: May 23, 2018]

[12] Aguiar DPO. Contribution to the study of the dam safety index - DSI (Contribuição ao estudo do índice de segurança de barragens – ISB). [thesis]. Campinas: Universidade Estadual de Campinas; 2014

[13] Medeiros CHAC, Dias GG, Anderáos A, Leonardi C, Pimentel CEB, Neumann C, Osako C, Coelho DP, Faria EF, Willrich FL, Patias J, Oliveira JÁ, Araújo LMN, Matos SF. Module I: DAMS - Legal, Technical and Socio-Environmental Aspects (Módulo I: BARRAGENS - Aspectos legais, técnicos e socioambientais) [Internet]. 2013. Available from: https://capacitacao.ana.gov.br/conhecerh/handle/ana/2179 [Accessed: September 10, 2018]

[14] Fusaro TC, Dias GG, Anderáos A, Leonardi C, Pimentel CEB, Neumann C, Osako C, Coelho DP, Faria EF, Willrich FL, Patias J, Oliveira JÁ, Araújo LMN, Matos SF. Module III: Dam Management and Performance (Módulo III: gestão e desempenho de barragens) [Internet]. 2012. Available from: https://capacitacao.ana.gov.br/conhecerh/handle/ana/2179 [Accessed: June 18, 2018]

[15] ANA – Agência Nacional de Águas e Saneamento Básico. Generation of Spots for Classifying Dams According to the Associated Potential Damage: Simplified methodology (Geração de manchas para classificação de barragens quanto ao dano potencial associado: Metodologia simplificada) [Internet]. 2017. Available from: https://www.snisb.gov.br/Entenda_Mais/capacitacao/Arquivos_Cursos/apresentacoes-do-curso-de-mancha-dpa/geracao-de-manchas-de-dpa-metodologia-simplificada.pdf [Accessed: June 18, 2018]

[16] ANA – Agência Nacional de Águas e Saneamento Básico. Resolution No 132 of 22 February 2016 (Resolução n° 132, de 22 de fevereiro de 2016) [Internet]. 2016. Available from: https://arquivos.ana.gov.br/resolucoes/2016/132-2016.pdf [Accessed: January 31, 2022]

Chapter 5

Uncertainty Factors Influencing Hydroelectric LCA Studies: A Review

Marla T.B. Geller and Anderson Alvarenga de Moura Meneses

Abstract

Despite the increase in research on Life Cycle Assessment (LCA) of Hydroelectric Power Plants (HPP) there are issues that need to be better discussed. This review aims to discuss factors that influence HPP LCAs such as: indirect emissions, different stages of HPPs (construction, operation, and decommissioning), scale/productivity of HPPs, types of projects (reservoir and run-of-river) and use of the ground. Most of the results obtained by HPP LCAs indicate that the construction phase is the most influential phase for indirect emissions due to the use of steel and concrete. The comparison of the HPP's LCA results with the LCA of other energy sources indicates that for the analyzed category Global Warming Potential (GWP), the HPPs present a good environmental performance considering the quantified emissions, their productivity and useful life. The present review highlights some uncertainty factors that influence HPP LCA studies and cites the need to carry out future studies on the environmental impacts of HPPs including these factors.

Keywords: hydropower plants, life cycle assessment, GHG emissions, indirect emissions, renewable energy

1. Introduction

A major challenge in today's world is to meet the demand for increased energy production considering environmental factors. The concern with sustainability in the sector leads countries to sign agreements for the replacement of energy production from non-renewable sources (coal, natural gas, oil and derivatives, uranium) by renewable ones (charcoal, hydraulic, wind, solar photovoltaic, biomass). Brazil has a privileged position in terms of energy production from renewable sources. According to the report [1] the share of renewables sources in the Brazilian electricity matrix reached 84.8% in 2020, against 23% in the rest of the world and 27% in OECD countries. In Brazil, hydraulic energy is positioned as the main generating source, where 67% of the energy generated in 2021 comes from hydroelectric plants [2]. Although Brazil occupies a privileged position in the renewable energy sector, in 2020 Brazil lost its place to China, which added 12.6 GW of hydroelectric capacity and regained its leadership. Many HPPs are being implemented, especially in China, India, and in northern Brazil, with the objective of increasing renewable energy production and meeting the growth of demand. Considering the above, it is important to assess the environmental impact caused by hydroelectric plants, since large hydroelectric plants are part of the energy matrix scenario in Brazil and in the world [3].

There are many factors that influence the analysis of HPP emissions, which is why it becomes a complex task. One of the ways to carry out this analysis, considering these factors, is through the Life Cycle Analysis (LCA). LCA is a methodology that makes it possible to analyze environmental damage and is defined in ISO [4] as: the study of environmental impacts throughout the life of a product from obtaining the raw material, passing through production, use and ending with discard it. LCA methodology approach holistically the entire process, identifying the most significant impacts, pointing out improvements and at what stages they can be applied. In this way, it prevents damage from spreading along the stages, causing a chain effect from one environmental problem to another or from one region to another [5]. According to ISO [4], the methodology for LCA comprises four steps.

1st) Objective and scope definition: the objective and scope must be clearly defined to ensure that no relevant part is omitted.

2nd) Life Cycle Inventory (LCI): in this step, data collection and calculation procedures are carried out to quantify the inputs and outputs relevant to the study of the system, as defined by the objectives and scope in the previous step.

3rd) Impact Analysis (LCIA): the third phase of the LCA aims to assess, quantify, and convert the environmental loads caused by the inputs (raw materials and energy) and outputs (waste and other emissions) of the system, into impacts on health, the environment, and the use of natural resources. Mandatory activities at this stage are the selection of impact categories, the definition of category indicators and characterization models, in addition to the classification and characterization of data. An impact category is a class representing environmental problems, such as global warming, acidification, human toxicity, etc.

4th) Interpretation: the last phase of the LCA aims to identify the significant environmental issues present in the results of the previous phases; evaluate the methodology, check the consistency, completeness and sensitivity of the data and propose recommendations for improvements in the performance of the system.

The LCA methodology is standardized [5]. Currently, LCA is being used for decision making in choosing the best option between products and processes in many contexts such as chemical engineering [6], in the use of disposable packaging [7], in agriculture [8], in the transport of products [9], in building construction [10], and frequently in the production of energy.

Therefore, analysis of emissions in energy production, through the LCA methodology, can be carried out in different contexts, such as: In Queiroz et al. [11] LCA was carried out in the process of producing biofuel from a palm tree in the Brazilian Amazon. Matuszewska [12] identified the configuration of geothermal systems using LCA. References [13–19] applied LCA to study the environmental damage of photovoltaic systems. Brizmouhun et al. [20] carried out the LCA study with the aim of identifying the environmental damage of power generation in Mauritius. Puettmann and Wilson [21] compared the cradle-to-gate total energy and major emissions for the extraction of raw materials, production, and transportation in wood products industry in United States. In Piasecka et al. [22] was used LCA to analyze and compare the environmental impact of offshore and onshore wind farms, considering 2 MW installations, which are the most frequent in Central and Eastern Europe.

In relation to HPPs, the analysis of emissions produced by the LCA methodology is accompanied by many uncertainties. Each plant has specific characteristics, such as location, size, type, productivity, land use, among others. All this influence the analysis causing a need for a study with well-defined methodology. HPP are becoming the solution to the problem of energy demand in many countries in the world, justifying the importance of using LCA to analyze the environmental impacts caused by this type of plant [23].

We can cite some examples such as: Pang et al. [24] compared the environmental impacts of a small HPP in China with similar ones located in other parts of the world, while Suwanit and Gheewala [25] evaluated the environmental impacts through the LCA of mini HPPs that generate electricity in Thailand. The biggest energy producer in the Nordic countries is Vattenfall AB. Vattenfall's main markets are Sweden, Germany, the Netherlands, Denmark, and the United Kingdom. Vattenfall [26] presents the LCA study carried out by Vattenfall, evaluating the environmental impacts of different energy sources in the Nordic countries. Santoyo-Castelazo et al. [27] analyzed the environmental impacts of HPPs in Mexico and other energy sources that are part of the supply network and presented a comparison between them. Flur and Frischknech [28] conducted a study which had the objective of describing the environmental impacts of construction, operation and decommission of HPPs with the focus on Switzerland, extrapolated to other regions such as Brazil. Hanafi and Riman [29] evaluated life cycle of a mini HPP in Simalungun - Indonesia and showed that the most evident impacts are carcinogenic, and eco-toxicity in marine and freshwater biota generated from the construction of the mini HPP.

After this preliminary study, we observed that there are extensive possibilities of applications for LCA methodology. An approach to the most influential aspects for analysis using LCA is the objective of this work, and we consider it a contribution to the current state of the art. The analyzed aspects include: the different phases of the LCA, the importance of indirect emissions in studies of hydropower reservoirs, the scale/capacity ratio, land use. We added a discussion on the type of run-of-river hydropower and the challenges of using more sustainable technologies.

The present work demonstrates the importance of the LCA methodology when considering all phases of the life cycle of HPP and highlights specific characteristics of HPPs when their environmental impacts are analyzed. Brazil is a developing country, and, like other similar countries, it has an unexplored hydraulic potential. HPPs can become one of the best solutions to meet energy growth and demand. However, the challenge for these countries is to study and analyze the best way to produce energy without causing major impacts on the environment. And the LCA methodology has been an aid tool for such studies.

2. Methodology and materials

The methodology used in this paper included six steps as in Geller et al. [23]:

i. Definition of the objective and delimitation of the research scope: The objective of the research is to present some characteristics of the HPPs that influence the analysis of environmental impacts through the LCA. This review includes comparative analyzes of the LCA of HPPs and another energy sources.

ii. Selection of sources in the literature: The selection of research sources was conducted with the following criteria: 1st) LCA reports published between 2000 and 2021, containing analysis of the environmental impact of HPPs and other energy sources; 2nd) studies that identified influencing factors on the LCA of HPPs. The search was carried out using the keywords: LCA, Hydropower, GWP, emissions, following this order of priority.

iii. Definition of factors that most influence HPP's LCA studies: There are many factors that cause uncertainties and influence LCA studies for hydroelectric plants. To carry out this study, the following factors were selected: Indirect emissions, types of HPPs (reservoir and run-of-river), the different phases of the life cycle, land use, location, scala/productivity.

iv. Review including only HPP's LCA: A review of articles related to HPP's's LCA was necessary to compare the different contexts, highlighting a study carried out in a HPP's in northern Brazil. The impact category highlighted was the Global Warming Potential – GWP, which according to Acero et al. [30] "expresses climate change referring to the global temperature caused by greenhouse gases released by human activity, measured in the reference unit kg of CO_2 equivalent (kg CO_2 eq)".

Countries like China, Brazil, USA, Canada, and Russia have a high production of electric energy through HPPs [31]. However, there are few studies on their emissions analyzed through LCA, leading to the need to include here some reviews carried out by other authors. With the results produced by other works cited in this review, it was possible to make comparisons between hydroelectric plants with different characteristics (in Section 4) and between hydroelectric plants and other energy sources (in Section 3).

v. Recognition of HPP LCA uncertainties: Compiling the results of the reviewed works and recognizing the uncertainties related to HPP LCA constitutes a significant step towards the objective of the study.

vi. Identification of challenges for future research: to highlighting points to be better studied and which constitute challenges for future research is one of the characteristics of the review study. The conclusion includes the authors' view on this challenge.

3. A comparison between hydropower and other sources using LCA

It is known that different energy sources have their emissions influenced by their characteristics. To recognize the environmental feasibility and sustainable character of each one, it is necessary to carry out a quantitative comparison of these emissions. And one way to make this comparison is through results from LCA. This section also includes a comparison with the LCA study carried out by the authors at the Curuá-Una HPP, located in the Amazon in northern Brazil.

The electricity and heat distribution company Vattenfall presented in Vattenfall [26] the LCA data for nuclear, hydro, wind, solar and biomass energy. Vattenffal is one of the biggest producers of electricity and heat in Europe. Its most important markets are Sweden, Germany, Holland, Denmark and the United Kingdom. The methodology used pela Vattenfall divided the life cycle into 4 stages, namely: (i) production and transport of fuel; (ii) plant operations; (iii) infrastructure that includes construction, maintenance and decommissioning of the plant and (iv) radioactive waste management. **Table 1** presents the main energy production technologies and their respective contributions. According to Vattenfall [26], the largest amount of emissions is generated by the coal plant in the operational phase. This reinforces the conclusion of the analyzes that the emissions from the construction phase are higher for plants that do not burn fuel but use renewable resources such as HPPs and wind farms. For biomass and coal-burning fuel plants, emissions are highest in the operational phase [39].

DOI: http://dx.doi.org/10.5772/intechopen.1000185

SOURCE	Coal	Oil	Natural gas	Biomass	Solar PV	Wind	Nuclear	Hydro	
				($kgCO_2$ eq/MWh)					
[32]	1230	1213.4	855	97.3	76.3	46.4	17.1	13.2	China (Ecoinvent)
[26]						15.6	4.44	7.26	Nordic Countries (Electricity mix)
[33]	600–1050	530–900	380–1000	8.5–120	13–190	3.0–41	3.0–35	2.0–20	Literature review
[20]	1444	754		29				8.6	Mauritius (Electricity mix)
[31]	888	733	499	14.0–650	9.0–300	8.0–124	24.2	2.0–75	Literature review
[34]	1118		514			16.9–30.4	39.5	27.7–43.8	Literature review
[35]	960–1050	778	443	118	13	9	15	2.0–15	Literature review
[36]	975.2	742.1			53	29.5	24	11.3	Japan
[37]	900–1200	790–900	400–500			4.6–55.4		0.2–152	Literature review
[38]						28.6 ± 3.2	12.4 ± 1.5	3.5 ± 0.4	China

Table 1.
Life cycle GHG emissions ($kgCO_2$ eq/MWh) of different technologies.

A literature review with 167 LCA studies was performed by Turconi et al. [33]. GHG emissions have been identified for the most diverse energy sources, including coal, lignite, natural gas, oil, nuclear, biomass, hydro, solar PV and wind. **Table 1** presents the results for each technology and points out that coal, lignite, natural gas and oil have the greatest impact compared to hydro, nuclear and wind power.

In the Republic of Mauritius, the main source of energy is fossil fuels and Brizmouhun et al. [20] used the LCA methodology to compare GHG emissions from these plants and eight other hydroelectric plants, of which 4 are reservoir and 4 are run-of-river plants. The results presented in **Table 1** reinforce the higher GHG emission for fossil fuel plants compared to HPPs.

In the work of Gagnon et al. [35] several energy sources were analyzed with the LCA methodology. A point to emphasize in this study is that the results presented the lowest impact for run-of-river plants followed by wind, solar photovoltaic and nuclear plants. The HPPs with reservoir had the worst performance.

The environmental impacts of electricity generation in China were analyzed by Feng et al. [32] through LCA, using data from eight different technologies. For the analysis of energy that has oil, natural gas, hydro, and photovoltaic energy as sources, the Ecoinvent base processes (repository with more than 18,000 datasets) were used. To analyze the impacts produced by nuclear, coal, biomass and wind energy, the studied plants are in China [18]. The study shows that CO_2 emissions from fossil fuel-based technologies are much higher than emissions from renewable sources of energy when analyzed over the life cycle (**Table 1**).

The review of Amponsah et al. [31] includes 79 studies of LCA. The objective was to compare renewable energy sources (RETs) such as photovoltaic, wind, biomass, wave energy and hydropower with conventional energy sources such as oil, lignite, and coal (**Table 1**).

The research carried out by Hondo [36] pointed out that one of the factors that can reduce the impacts on energy production is the choice of technology, such as the material used in photovoltaic panels. The author proved with his study there was a reduction from 29.5 CO_2-eq/MWh to 20.3 CO_2-eq/MWh for wind energy and 53.4 CO_2-eq/MWh to 26 CO_2-eq/MWh for PV (**Table 1**). For this analysis, it measured the emissions of nine types of energy production technology, among them: nuclear, hydroelectric, geothermal, wind, solar-photovoltaic (PV)coal, oil, liquefied natural gas (LNG) and LNG combined cycle.

Littlefield et al. [34] included the environmental profile in their research on energy production feasibility analysis. The environmental profile uses LCA and assesses resource consumption, emissions to water and air, solid waste, and land use. Seven technologies were analyzed, among them: hydroelectric, wind, nuclear, natural gas, coal and biomass co-burning, geothermal and solar thermal resources. The results pointed to the wind farm with the best environmental performance (**Table 1**).

In Like et al. [38] wind, nuclear and hydraulic plants were evaluated with LCA, considering all stages of the life cycle. **Table 1** shows that wind power has a greater environmental impact than nuclear and hydro power. Considering the global warming potential, wind energy produces 28.6 ± 3.2 g CO_2-eq/kWh of GWP100 throughout its life cycle, which is higher than nuclear energy (12.4 ± 1.5 g CO_2 -eq/kWh) and hydroelectric (3.5 ± 0.4 g CO_2-eq/kWh).

We can see that the results presented in **Table 1** show the different approaches in the use of LCA to measure the environmental impacts on energy production. Knowing the characteristics of these plants that influence the different results constitutes a relevant study to aid in decision making when seeking sustainability.

The LCA methodology used to compare various energy sources requires a standard of functional unit. This allows the productivity of each energy source to be considered. In the case of the results presented in **Table 1**, the functional unit is $kgCO_2$-eq/MWh. This standard is important because, like hydroelectric plants, which feature high productivity and longer lifespan than other technologies, their environmental impact is diluted both in the amount of energy produced and in its lifespan. In other words, to produce the same amount of energy a wind farm needs many resources (panels, batteries, inverters) that use raw materials (inputs such as metals and energy), contributing to the various categories of environmental impact [16].

Table 1 shows that the greatest variation in GHG emission quantification is for HPPs (0.2–152 $kgCO_2$-eq / MWh) and biomass (8.5–650 $kgCO_2$-eq / MWh), a result discussed in Topic 5.6.

4. LCA of HPPs: comparing different characteristics

To further complement the study an analysis of LCAs limited to HPPs with different characteristics was conducted including different phases of LCA, size of plants, type of plants (reservoir or run-of-river) and land use.

When comparing LCAs done for different HPPs, to obtain better interpretation and analysis, it is very important to consider the objective of each study as this leads to specific results. Some studies consider only the GWP factor [16, 27, 40], whereas others present the total of emission for some factors [37, 41]. There are studies that review several LCAs done by different authors and compare them, such as the study by Refs. [33, 37]. Research in Refs. [24, 39] evaluated the most representative impact categories for each stage of LCA applied to hydropower. The review in this session considered analyzes only of HPPs with LCA methodology, independent of other sources of energy production. All were standardized with the functional unit kWh or MWh, and different results for these analyzes are shown in **Table 2**.

Varun and Bhat [40] studied emissions from 6 hydropower plants located in India, of which three are canal-based projects and three are dam-based projects. The results of this study, in relation to GHG emissions, are shown in **Table 2**. The specific characteristics of each HPP such as capacity, type of technology, location and head size produce variation in the values found in the analysis performed with LCA.

Five run-of-river mini HPPs, located in Thailand, were the object of study in Suwanit and Gheewala [25]. Considered useful life was 50 years and functional unit was 1MWh. For analysis of the results, the averages of the five mini HPPs were calculated and are shown in **Table 2**. Reproducing the authors' conclusion "the main contributors to the impacts are the materials used in construction such as gravel, sand, cement, steel, iron, copper and the energy used by the equipment" [25].

A review carried out by Raadal et al. [37] compared the emissions caused by 28 reservoir-type HPPs and 11 run-of-river HPPs, as shown in **Table 2**. Among the evidence of the result is the importance of considering the flooded area as a significant factor impact on the environment. The reservoir filling stage is highlighted as the biggest contributor to GHG emissions, even surpassing the construction stage.

In Hanafi and Riman [29] the LCA methodology was used for the analysis and a HPP located in Simalungun in Indonesia. The system operates with two generators, it is of the run-of-river type, has a capacity of 9 MW and productivity of 8MWh, registering 80% efficiency. Considering a useful life of 50 years, the main stages defined by the methodology were: pre-construction, construction, and operation. This study reinforced what other

	Total (kg CO_2-eq/MWh)	Study Site/Type	Type
[24]	28.4	China	dam-toe-based
[40]	32.23–35.35	India	canal based
[40]	11.91–31.2	India	dam-toe-based
[25]	17.62	Thailand	run-of-river
[29]	1.2	Indonesia	run-of-river
[42]	52.7	Thailand	run-of-river
[43]	5.43–8.93	UK	run-of-river
[44]	4.33	Brazil	reservoir
[45]	13.06–19.12	India	run-of-river
[46]	21–40.63	Literature review	reservoir
[46]	3.0–47	Literature review	run-of-river
[46]	256.63	Literature review	pumped storage
[33]	2	Literature review	run-of-river
[33]	15	Literature review	reservoir
[37] *	4.0–152	Literature Review	reservoir
[37] **	0.2–11.2	Literature Review	reservoir
[37]	4.9	Literature Review	run-of-river
[47]	>150.0	Literature Review	reservoir
[47]	4.0–14.0	Literature review	run-of-river

**Including gross emissions from flooded land.*
***Excluding emissions from flooded land.*

Table 2.
HPP- life cycle GHG emissions (kg CO_2-eq/MWh).

results have also shown, the greatest impact to the construction phase. However, the most influential categories for this phase are ecotoxicity and the release of carcinogenic materials. The authors justify this result using materials such as steel, nickel, and concrete for the construction of the duct. They emphasize that the amount of GHG emission (1.2 kg CO_2-eq/MWh) is lower than the categories mentioned above (**Table 2**).

The review carried out by Kumar et al. [47] presents estimates of GHG emissions from reservoir and run-of-river plants, reporting values that exceed 150 kgCO_2-eq/MWh for the former and a range of 4–14 kg CO_2-eq/MWh for the last. The authors emphasize that this different behavior in terms of emissions is due to the use of land by reservoir-type HPPs and that all phases can contribute to this, due to the emission of methane with the decomposition of vegetation (construction and operation phases) and the sludge deposited throughout the plant's useful life (decommissioning phase).

The study by Pascale et al. [42] compared the emissions produced by a small dam in a rural community in Thailand with the emissions from larger dams, with the objective of finding the best alternative for the electrification of these communities. For this, it carried out the LCA of a small 3 kW system considering its entire life cycle, that is, from the cradle to the grave. The results presented in **Table 2** reflect the trend found in the literature, that is, larger HPPs, due to their higher productivity, have a lower environmental impact per kWh produced, compared to smaller HPPs.

Itaipu HPP plant is a binational power plant (Brazil-Paraguay), located on the Paraná River, on the border between these two countries. Until 2012 it was considered the largest hydroelectric plant in the world, with a capacity of 14,000 MW. Ribeiro and Silva [44] used the LCA methodology to make an inventory of material and energy consumption, atmospheric emissions and land use and transformation. According to the authors, the good environmental performance (4.33 kg CO_2-eq/MWh) obtained as a result of the study (**Table 2**) is justified by the high productivity of this large hydroelectric plant.

Three run-of-river HPPs located in the United Kingdom (UK) were analyzed in Gallagher et al. [43]. Each one had a capacity of 650 kW, 100 kW and 50 kW and the result of the study shows emissions of 5.43 kg CO_2-eq/MWh, 7.39 kg CO_2-eq/MWh and 8.93 kg CO_2-eq/MWh respectively (**Table 2**). According to the authors, there are few LCA studies for small (~100–1000 kW) and micro (~10-100 kW) HPPs that highlighted the importance of knowing their emissions because hydroelectric plants have significant growth in some regions (**Table 2**).

Kadiyala et al. [46] created an index categorizing HPPs according to capacity (micro, small and large) and type (impoundment, diversion, pumped storage, miscellaneous hydropower works). The mean GHG emission resulting from small HPP dams was higher than large hydropower dams of the same type. The highest average emissions were found for pumped storage (**Table 2**).

Pant et al. [45] selected three small run-of-river hydropower plants located in India with capacities of 30 MW, 33.33 MW and 51 MW. The analysis was carried out using the LCA methodology, considering, in addition to the capacity, the head of the power plants. The results show that emissions are lower when there is higher productivity, on the other hand, they increase with the increase in the head of the plant.

4.1 LCA of HPPs detailed in different phases

Different stages of the life cycle of a HPP produce emissions of the most diverse natures and quantities. In this session, we present some studies on HPP LCA found in the literature that confirm this statement.

Pang et al. [24] confirmed in their research (**Table 3**) using the LCA methodology, that the construction phase in the life cycle of a HPP is the one with the greatest environmental impact due to the emission of GHGs. They analyzed the Guanyinyan power plant, in northeast China, which has 2 turbines, a capacity of 1.6 MW and an average annual production of 6.28GWh. They used the 1 MWh functional unit and considered the plant's useful life to be 30 years. To complement the research, they carried out a sensitivity analysis, increasing and decreasing inputs by 10%, and studying the consumption of materials such as steel, cement, and energy. They obtained ±6.8% in the results and concluded that "it is necessary to optimize structural designs using new materials and best practices to reduce the level of emissions in relation to the energy generated [24]."

The LCA study carried out in Geller and Meneses [39] specified the emissions from the Curuá-Una hydroelectric plant located in the Amazon region in its different phases (**Table 3**). Zhou [48] included in his study the HPP of Nam Theun 2, located in southeastern Laos with a capacity of 1070 MW (**Table 3**), also in different phases.

	Total	Construction	Operation	Transportation	Maintenance	Disposal
	(kg CO_2-eq/MWh)					
[24]	28.4	27.3		0.2	0.9	0.0
[39]	5.4	4.8	0.11	0.46		0.0
[48]	78.1	2.5 ± 0.5	75 ± 0.5			0.55–0.65

Table 3.
HPP - life cycle GHG emissions (kg CO_2-eq/MWh) – Phases.

In Zhou [48] they pointed out that the operational phase of the plant is the phase had the greatest impact due to the emissions produced by the reservoir. The gases carbon dioxide is produced by aerobic decomposition and methane by anaerobic decomposition and are the gases produced in greater quantity in the plants located in tropical regions. Methane accounts for 85% of total emissions at warmer temperatures.

These three studies are used here because they relate a high amplitude in the results that can be attributed to the different forms of application of the LCA methods.

In Pang et al. [24] the LCA analysis of a small HPP (1.6 MW) with a useful life of 30 years was performed. In the other case Geller and Meneses [39] the HPP has a capacity of (30.3 MW) and the analysis was performed for a useful life of 100 years. The results as shown in **Table 3** were quite different, justified by the fact that the greater capacity and longer useful life results in greater productivity, causing the quantities to be diluted in relation to the production factor (emission/production factor). And as already mentioned, the high value of the operation phase in Zhou [48] is due to the inclusion of direct emissions of methane and carbon dioxide gases, produced by the decomposition of biomass in the flooded area of the reservoir. It is worth mentioning that in Refs. [24, 39] only indirect emissions were analyzed.

Geller and Meneses [39] investigated the environmental impacts of the construction, operation and deactivation phases of the HPP Curuá-Una plant located 70 km from the city of Santarém, in the northern region of Brazil, that is, in the Brazilian Amazon. The HPP was inaugurated on 08/19/1977 and at the time of the study (2016), the plant operated with three turbines with an operating capacity of 30.3 MW and with an efficiency rate of 92.89%, being considered by Brazilian standards as a large plant. The investigation was carried out with real data and data from the Ecoinvent database and the categories analyzed include:

- Global Warming Potential (GWP),
- Acidification Potential (AP),
- Abiotic Depletion Resources (ADP),
- Freshwater Aquatic Ecotoxicity Potential (FAETP)
- Human Toxicity Potential (HTP),

Impact category	Reference unit	Complete life cycle	Construction	Operation	Transportation	Decommission
AP	[kg SO_2-eq]	0.0223	0.0189	0.0009	0.0025	−0.000
GWP 100y	[kg CO_2-eq]	5.4659	4.8922	0.1121	0.4679	−0.0065
ADP	[kg Sb eq]	0.0312	0.0247	0.0033	0.0032	0
FAETP 100y	[kg 1.4-DCB eq]	2.4505	2.2971	0.1169	0.0371	−0.0007
HTP 100y	[kg 1.4-DCB eq]	7.2858	6.4277	0.6267	0.2345	−0.0031

Table 4.
Contribution of Curuá-Una life cycle phases to each impact category [39].

The results presented in **Table 4**, show the four phases of the life cycle analyzed in Geller and Meneses [39]. Among the categories analyzed, the most affected are HTP, GWP and FAETP and the stage with the highest emission is the construction phase. Fossil energy was used in this phase, which is a major contributor to these impact categories. The construction phase has emissions related to all impact categories and among the main contributors to this fact is the use of concrete used in the dam and steel used in infrastructure and equipment such as generators and turbines. It is noteworthy that the methodology used did not include direct emissions, that must be measured directly in the reservoir, as already mentioned, so the low results for the operation phase with the appropriate emissions of CH_4 and CO_2 are noted.

All results presented confirm the need for specific studies in each context and for each objective, since the characteristics of each plant are different, and the objectives of the studies require adequacy of the LCA methodology. According to Geller and Meneses [39] "a direct comparison between HPPs is difficult and should be made carefully because HPPs are highly site-specific, and their environmental impacts are associated with their different characteristics."

5. Discussion

This session discusses issues regarding the specific characteristics of hydropower that impact the Life Cycle Analysis results when analyzing emissions. The diversity of studies and contexts in which this methodology is applied was observed in the previous sessions, generating variation in the results.

5.1 Direct and indirect emissions of HPPs

Emissions produced by the generation of energy are classified in two ways: indirect and direct [28]. The former are the emissions caused by the construction, implantation, and deactivation of the plant, which according to Steinhurst et al. [49] include infrastructure of roads and transmission lines, the work of implantation, manufacture of materials, transport, disposal of material, etc. According to Dones et al. [41], the largest contributor to GHG emissions in this category includes cement production and the use of diesel for electricity. Hanafi and Riman [29] state that "the biggest contributing factor related to the infrastructure for the emission of GHGs is the production of concrete and the transport of stones for the construction of dams and tunnels". Indirect emissions can represent less than 20% of generating plants using fossil fuel and more than 90% of generating plants using renewable and nuclear sources [49].

And direct emissions are those resulting from the phase in which the plant is in operation, such as burning of fuel used to operate the plant, use of land/flooded area, goods and services for plant operation, etc. In relation to the direct emission produced by the hydroelectric plants, the decomposition of the biomass of the flooded soil of the reservoir is one of the most significant [29]. The impacts caused by these emissions are classified according to the scope of the area they affect in global impact, regional impact and local impact, that is, some may have a significant impact in one region, but not in another. For example, global warming, depletion of the ozone layer, biotic and abiotic resources caused by emissions from a particular location, can cause global impacts; land use will be responsible for local impacts [4].

Hydroelectric plant is a renewable energy source that is not entirely clean, because if considered throughout its life cycle, it has direct and indirect contributions to the production of GHGs. Some studies in Refs. [28, 49–52] are concerned to show that the idea that hydroelectric plants have low levels of emissions is somehow wrong. They argue that emissions can sometimes exceed those of plants that use non-renewable raw materials, such as fossil fuels, in their implementation phase and in some cases also in the operational phase, as described by Refs. [53, 54]. The review presented in this chapter is not intended to provide details of these results.

5.2 Reservoir HPPs and indirect emissions

The consumption of concrete, steel, energy, fuel, and other materials for the construction of the plat, in addition to the use and transformation of the soil, are the most responsible for the indirect emissions that impact the construction phase.

Geller and Meneses [39] state that the construction stage at the Curuá-Una HPP was the most critical phase in relation to the indirect emissions caused (**Table 3**), due to the inputs for deploying the HPP, which is corroborated by other authors [24, 41, 49, 55]. Pang et al. [24] concluded that steel and concrete are the largest contributors among the materials used as inputs due to their production chains.

References [37, 41] reinforce that cement production, the use of diesel for electricity and the transport of stone for construction of dams and tunnels are major contributors to GHG emissions. 90% of emissions from renewable plants come from the construction phase [49]. Vattenfall [26] say that plants that use renewable resources for energy production (hydro, solar, wind), that is, do not burn fossil fuel, have their most impacting indirect emissions in the construction phase.

5.3 Scale/capacity of the hydropower plants

According to Refs. [24, 35, 41, 44] the capacity and scale of projects also influence GHG emissions, stating that smaller systems have worse environmental performance than systems with greater capacity. This is since the higher productivity of the latter, reduces the ratio between the emission and the energy produced in MWh during the useful life of the plant. Zhang et al. [56] carried out LCA of two HPPs in China, with 44 MW and 3600 MW, and the GHG emissions were 44 and 6 kg CO_2-eq/MWh, respectively.

We can explain by observing the statements of these authors that the ratio between the impact per MWh and the higher productivity of the plant, considering its long-life cycle, are factors that can make the HPPs the more environmentally viable option to meet greater demands, because it dilutes the amount of impacts over its useful life. However, the need to serve small populations or rural communities located at great distances from large HPPss, small plants are still a very viable alternative for energy production, having better environmental performance than fossil fuel plants, that is often used in this context.

Environmental impacts are influenced by several factors in addition to the productivity of the plant, as already seen. Therefore, the analysis of the LCA of HPPs with different scale and capacity must be carried out with great care, even when the values are parameterized.

5.4 Land use and land use change (LULUC)

A highly discussed issue with respect to LCA of HPPs is to land use change and Land Use Change (LULUC). Land use of HPPs implies land transformation (flood area and implantation) and occupation (entire occupied area concerning the time of use). According to Dorber et al. [57], hydroelectric pumped storage and storage hydroelectric plants, which use dams to store water in reservoirs to allow flexible production of electricity, cause LULUC. In addition to filling the reservoir, also due infrastructure, including power lines and access roads.

It should also be noted in Fthenakis and Kim [58] that state the occupancy factor is indirectly proportional to the time that a given area is used to generate renewable energy, that is, the longer the time, the lower the occupancy factor. On the other hand, for the production of energy from non-renewable sources (biomass, coal, natural gas, etc.), the relationship is direct, that is, the longer the time, the greater the occupancy factor.

The analysis of land use, such as occupation or transformation of HPPs using LCA is necessary. Take the example of Balbina HPP in Brazil. According to Fearnside [53], this plant will produce more GHG than fossil fuels because the proportion of the reservoir size per unit of energy generated is very high. The example of the LCA of the Curuá-Una plant in Geller and Meneses [39] in the Amazon region can illustrate these conclusions. The result of the analysis in this regard was 1.33E-02 km2.y/MWh. For an efficiency of 92.89%, its projected production is 29,976,720 MWh in 100 years. By reducing the operating time to 50 years considering the same efficiency, the HPP would produce half of that energy, and consequently would increase its occupancy factor and its environmental damage.

5.5 Others important factors that influence the emissions of HPPs

The environmental impacts produced by reservoir and run-of-river plants were presented by Gagnon and Vate [59]. For reservoir plants, the average was 15 kg of CO2-eq/MWh and for run-of-river plants, the average was 2 kg of CO2-eq/MWh. As already mentioned, steel and concrete are the main contributors to these emissions, as they are consumed in greater quantities in reservoir plants. The authors detailed the factors that contribute to emissions in the construction phase and the study includes the following criteria: (i) run-of-river hydroelectric plants versus reservoir plants; (ii) materials used (earth/rock versus reinforced concrete); (iii) volume of dikes and dams, which may be site-specific; and (iv) total size of the project or group of plants [59].

An important issue to consider when comparing run-of-river and reservoir plants is the reliability of the electricity supply. LCA studies of HPPs present results pointing to better performance for run-of-river HPPs [33, 37, 46, 47]. However, according to Gagnon and Vate [59], reliability is only possible when the energy source (water) can be stored in a reservoir, being available for operation. We emphasize that production reliability must also be considered when comparing reservoir HPP with plants that have intermittent sources, such as wind and solar photovoltaic. In the absence of these resources (wind and sunlight) it will be necessary to complement them to meet the demand, which is usually through fossil resources, greatly increasing the environmental impact.

Another issue to be considered is the use of reservoirs for other purposes, such as irrigation, flow control, flood control, storage of drinking water and even fishing. Currently, large reservoirs are planned and built for these purposes, such as the Three Gorges HPP in China, which uses stored water to control the flow available during the dry season, to facilitate navigation and to irrigate [32, 47]. On the other hand, there are cases, such as the Sardar Sarovar dam, Gurajat in India, where the main purpose of its construction was the irrigation of arable areas and the storage of water for domestic and industrial consumption, while the generation of hydroelectric energy was considered a side benefit, although more recently it has increased its importance [60]. The study of LCA in these cases becomes difficult because the importance of the reservoir as an energy producer is not fully known [35].

The type of technology used its manufacturing location and the distance of travel for the implantation in the plant is a significant factor for the environmental impacts [31]. Transporting this equipment from origin to destination consumes fossil fuel, leading to a significant increase in emissions. Like the Curuá-Una HPP, in Brazil, in which most of the equipment was manufactured far from the implantation site and it was often necessary to transport turbines and generators by river and road [39].

5.6 LCA of HPPs uncertainties and future challenges

Some uncertainty factors in LCA of HPPs were listed in this study. In general, it can be attributed to the lack of a standard for applying the methodology. However, the complexity and diversity of these systems makes this standardization difficult.

Initially, the use of a standard functional unit in the comparison of the different plants is not enough to guarantee a correct analysis of the results. There is divergence in the results when we have variation in the limits or boundaries of the system. Is important know well which specificities of each study will be included in the inventory.

For example, the LCA methodology applied to plants that have a reservoir most of the time includes all phases of the life cycle, namely, construction, operation/maintenance and decommissioning. However, there are few reports of LCAs that include the measurement of methane and carbon dioxide emissions that occur during the filling of the reservoir or the flooded area during the operational phase. There are many factors that can influence these measurements, such as pre-existing vegetation, location and climate, size and depth of the reservoir, among others. It is noteworthy that these emissions can significantly influence the results [52].

According to Littlefield et al. [34] the uncertainty factors are greater in the production of energy from renewable sources and Turconi et al. [33] add that among these, biomass and hydroelectric technologies are the most divergent in the LCA results (**Tables 1** and **3**). Temperature variation is also an important factor, according to the authors [33, 34], and reservoirs in tropical regions may be subject to higher GHG emission due to high temperature, which accelerates the biomass decomposition process.

Considering the above, there is a need for more specific conduct for each case of HPP LCA. This constitutes a great challenge. As a contribution to facilitating this process, we include the following recommendations: (i) elaboration of a framework to assess uncertainties; (ii) definition of a standard to classify dams according to their size, geographic location and climate; (iii) a framework to conduct the collection of essential data for hydroelectric plants, defining the system boundary in the inventory phase; (iv) construction of a common vocabulary, a practical application guide, so that researchers from other areas can contribute to the application of the LCA of HPPs methodology, with the objective of reducing these uncertainty factors.

The conclusions lead us to enumerate some alternatives to reduce the environmental impacts in energy production from HPPs: (i) future research to quantify emissions from hydroelectric plants should be conducted to include their entire life cycle, from cradle to grave, considering direct and indirect emissions; (ii) using LCA to analyze emissions from large hydroelectric plants located in developing countries such as China, Brazil and India can bring elements to a better choice of energy production in very extensive territories; (iii) the integration between different technologies of renewable sources, in an optimized way, can be a good alternative to meet the growing demand considering the parameters of sustainability; (iv) choosing more modern technologies for infrastructure (turbines, dams, pipelines, etc.) can reduce emissions during the construction phase of hydroelectric plants.

We understand that the optimized integration of different technologies in energy production is a viable way to meet the demand and, in this context, the hydroelectric plant plays an important role. However, the challenges of meeting the economic, social, and environmental viability of these projects have caused many concerns, discussions, and controversies.

Some research considers hydropower to be a clean, renewable energy source, while others claim that its emissions can sometimes be even higher than those of fossil fuels. It is important that this topic is discussed in a context based on the results of peer-reviewed scientific studies, such a scientifically based discussion will allow the rigorous evaluation of the most viable forms of energy production that attend the principles of sustainability.

6. Conclusion

The chapter presents a review of the HPP's LCA and compiles aspects that may influence these analyses, therefore the main contribution of this work is to highlight the different characteristics of the HPPs that impact the analysis of their indirect emissions.

Among the factors mentioned are the capacity and scale of projects, pointed larger systems have better environmental performance than smaller systems, due the ratio between the emission rate and the energy produced in MWh in the plant's lifetime.

Regarding the type if is reservoir or run-of-river, there is a factor that must be considered. Although reservoir plants produce more emissions, authors cite the use of these reservoirs for other purposes (water flow control, fishing, etc.), bringing additional benefits.

Land use must be critically analyzed when dealing with HPPs that use dams, due to the construction of its infrastructure including power lines and access roads. Although, the longer time a certain area is used for generation of renewable energy the lower is the occupation factor.

It is notorious and research shows that HPPs produce less emissions per energy generated than technologies that use fossil fuels. However, to evaluate the results among only renewables sources, such as wind, solar and hydroelectric, it is necessary to consider the intermittence of sources, that is, the constant availability of the natural resource (wind, water, solar) to generate energy. And this factor corroborates to the better performance of HPP in some cases.

Regarding the phases analyzed by the HPP's LCA, most studies state that the construction phase is the one that produces the most emissions in relation to renewable plants. But some studies also cite emissions from the flooded area and submerged vegetation as an important emission factor.

The results obtained by the research with LCA methodology exemplified here confirm that it can be widely used for the environmental analysis of electric energy production systems. However, the lack of standards for the preparation of inventories and for their analysis is pointed out by many authors as one of the difficulties for the use of LCA. The authors list in the challenges section some suggestions to minimize this problem.

In this context, it is considered that the more studies of the environmental impacts produced by the indirect emissions of HPS, using LCA, the more knowledge about the adequacy of the use of this methodology, the uncertainty factors and their restrictions will be consolidated, thus collaborating in the decision process involving HPPs.

Acknowledgements

The authors acknowledge the support of the Foundation for Research Support of the State of Pará (FAPESPA).

Author details

Marla T.B. Geller[1*] and Anderson Alvarenga de Moura Meneses[2]

1 Lutheran University Center of Santarém and Laboratory of Computational Intelligence – UFOPA, Santarém, Brazil

2 Federal University of Western Pará, Institute of Geosciences and Engineering and Laboratory of Computational Intelligence – UFOPA, Santarém, Brazil

*Address all correspondence to: marla.geller@gmail.com

References

[1] EPE - Empresa de Pesquisa Energética. Relatório Síntese Do Balanço Energético Nacional 2021: Ano Base 2020. Institucional. Rio de Janeiro: MME/EPE, 2021. Available from: https://www.epe.gov.br/pt/publicacoes-dados-abertos/publicacoes/balanco-energetico-nacional-2021. [Accessed: January 21, 2022]

[2] ANEEL. Agência Nacional de Energia Elétrica. Institucional. Available from: www.aneel.gov.br. [Accessed: February 12, 2022]

[3] REN21. Renewables 2021 - Global Status Report. Paris: REN21 Secretariat. Available from: www.ren21.net/wp-content/uploads/2019/05/GSR2021_Full_Report.pdf. [Accessed: March 10, 2022]

[4] ISO. Environmental Management - Life Cycle Assessment - Principles and Framework. Switzerland: ISO; 1997. Available from: www.iso.org/standard/37456.html. [Accessed: 2018-10]

[5] Azapagic A. Life cycle assessment and its application to process selection, design and optimisation. Chemical Engineering Journal. 1999;**73**:1-21. DOI: 10.1016/S1385-8947(99)00042-X

[6] Yue D, Kim A, You F. Design of sustainable product systems and supply chains with life cycle optimization based on functional unit: General modeling framework, mixed-integer nonlinear programming algorithms and case study on hydrocarbon biofuels. ACS Sustainable Chemistry Engineering. 2013;**1**:1003-1014

[7] Foolmaun RK, Ramjeawon T. Life cycle assessment (LCA) of PET bottles and comparative LCA of three disposal options in Mauritius. International Journal of Environment and Waste Management. 2008;**2**:125-138

[8] Meinsterling K, Samaras C, Schweizer V. Decisions to reduce greenhouse gases from agriculture and product transport: LCA case study of organic and conventional wheat. Journal of Cleaner Production. 2009;**2**:222-230. DOI: 10.1016/j.jclepro.2008.04.009

[9] Spielmann M et al. Scenario modelling in prospective LCA of transport systems. Application of formative scenario analysis. The International Journal of Life Cycle Assessmen. 2005;**10**:325-335 DOI: 10.1065/lca2004.10.188

[10] Mithraratne N, Vale B. Life cycle analysis model for New Zealand houses. Building and Environment. 2004;**39**: 482-492. DOI: 10.1016/j.buildenv.2003.09.008. Available from: https://www.academia.edu/1972332/Life_cycle_analysis_model_for_New_Zealand_houses?from=cover_page. [Accessed: March 10, 2022]

[11] Queiroz A, França L, Ponte MX. The life cycle assessment of biodiesel from palm oil ("dendê") in the Amazon. Biomass and Bioenergy. 2012;**36**:50-59. DOI: 10.1016/j.biombioe.2011.10.007

[12] Matuszewska D. Environomic Optimal Design of Geothermal Energy Conversion System Using Life Cycle Assessment (M.Sc. Thesis). School for Renewable Energy Science, Iceland. 2011. Available from: <https://skemman.is/bitstream/1946/7749/1/Dominika_Matuszewska_16.02.2011.pdf>. [Accessed: September 12, 2018]

[13] García-Valverde R, Cherni JA, Urbina A. Life cycle analysis of organic photovoltaic technologies. Progress in Photovoltaics: Research and Applications. 2010;**18**:535-558. DOI: 10.1002/pip.967

[14] Martinopoulos G. Life cycle assessment of solar energy conversion systems in energetic retrofitted buildings.

Journal of Building Engineering. 2018;**20**:256-263. DOI: 10.1016/j. jobe.2018.07.027.2352-7102. Available from: https://www.sciencedirect.com/ science/article/pii S2352710217305326. [Accessed: January 12, 2022]

[15] Laleman R, Albretch J, Dewull J. Life cycle analysis to estimate the environmental impact of residential photovoltaic system in regions with a low solar irradiation. Renewable and Sustainable Energy Reviews. 2011;**15**:267-281. DOI: 10.1016/j.rser.2010.09.025

[16] Desideri U, Proietti S, Zepparelli F, Sdringola P, Bini S. Life cycle assessment of a ground-mounted 1,778 kW photovoltaic plant and comparison with traditional energy production systems. Applied Energy. 2012;**97**:930-943. DOI: 10.1016/j.apenergy.2012.01.055

[17] Schwartfeger L, Miller A. Environmental Aspects of Photovoltaic Solar Power. Presented at the EEA Conference & Exhibition. New Zeland: University of Canterbury; 2015. Available from: <http://hdl.handle. net/10092/ 11214>. [Accessed: October 10, 2018]

[18] Frischknecht R, Itten R, Wyss F. Life Cycle Assessment of Future Photovoltaic Electricity Production from Residential-Scale Systems Operated in Europe (IEA-PVPS Task 12, Subtask 2.0, No. IEA-PVPS T12-05:2015). Golden, CO, USA: Garvin Heath, National Renewable Energy Laboratory; 2015. Available from: <https:// nachhaltigwirtschaften.at/ resources/ iea_pdf/reports/iea_pvps_ task12_ report_2015_lca_of_future_pv_ electricity_ production.pdf>. [Accessed: October, 2018]

[19] Aristizábal AJ, Sierra DC, Hernandez JA. Life-cycle assessment applied to photovoltaic energy: A review. IOSR Journal of Electricals and Electronics Engineering. 2016;**11**:06-13. DOI: 10.9790/1676-1105010613

[20] Brizmouhun R, Ramjeawon T, Azapagic A. Life cycle assessment of electricity generation in Mauritius. Journal of Cleaner Production. 2015;**106**:565-575. DOI: 10.1016/j.jclepro.2014.11.033

[21] Puettmann ME, Wilson JB. Life-cycle analysis of wood products: Cradle-to-gate LCI of residential wood building materials. Wood and Fiber Science. 2005;**37**(Corrim Special Issue):18 – 29. Available from: www.wfs. swst.org/ index.php/wfs/article/ view/1268. [Accessed: February 20, 2022]

[22] Piasecka I, Tomporowski A, Flizikowski J, Kruszelnicka W, Kasner R, Mrozinski A. Life cycle analysis of ecological impacts of an offshore and a land-based wind power plant. Applied Sciences. 2019;**9**:231. DOI: 10.3390/ app9020231. Available from: www.mdpi. com/journal/applsci. [Accessed: January 23, 2022]

[23] Geller MTB, Bailao JL, de Lima Tostes ME, de Meneses AA. Indirect GHG emissions in hydropower plants: A review focused on the uncertainty factors in LCA studies. DMA Desenvolvimento e Meio Ambiente. 2020;**54**:500-5017. DOI: 10.5380/ dma.v54i0.68640 e-ISSN 2176-9109

[24] Pang M, Zhang L, Wan C, Liu G. Environmental life cycle assessment of a small hydropower plant in China. International Journal of Life Cycle Assessment. 2015;**20**:796-806. DOI: 10.1007/s11367-015-0878-7

[25] Suwanit W, Gheewala SH. Life cycle assessment of mini-hydropower plants in Thailand. International Journal of Life Cycle Assessment. 2011;**16**:849-858. DOI: 10.1007/s11367-011-0311-9. [Accessed: October, 2018]

[26] Vattenfall. Life Cycle Assessment for Vattenfall's Electricity. Stockholm, Sweden: Public company (publ); 2021.

Available from: https://group.vattenfall.com/siteassets/corporate/who-we-are/sustainability/doc/vattenfall-lca-brochure.pdf. [Accessed: March 10, 2022]

[27] Santoyo-Castelazo E, Gujba H, Azapagic A. Life cycle assessment of electricity generation in Mexico. Energy. 2011;**36**:1488-1499. DOI: 10.1016/j.energy.2011.01.018

[28] Flur K, Frischknech R. Life Cycle Inventories of Hydroelectric Power Generation. ESU - Services, Fair Consulting in Sustainability, Comissioned by Oeko Institute e.V. Alemanha: Comissioned by Oeko Insitute e. V; 2012. Available from: http:// esu-services.ch/fileadmin/download/ publicLCI/flury-2012-hydroelectric-power-generation.pdf. [Accessed: September 12, 2018]

[29] Hanafi J, Riman A. Life cycle assessment of a mini hydro power Plant in Indonesia: A case study in Karai River. Procedia CIRP Science Direct. 2015;**29**:449-449. DOI: 10.1016/j.procir.2015.02.160

[30] Acero AP, Rodríguez C, Ciroth A. LCIA Methods Impact Assessment Methods in Life Cycle Assessment and their Impact Categories. Berlin: Greendelta; 2015. Available from: www.openlca.org/wp-content/uploads/2015/11/LCIA-METHODS-v.1.5.4.pdf. [Accessed: September 12, 2018]

[31] Amponsah NY, Troldborg M, Kington B, Aalders I, Hough L. Greenhouse gas emissions from renewable energy sources: A review of lifecycle considerations. RSER. 2014;**39**:461-475. DOI: 10.1016/j.rser.2014.07.087

[32] Feng K, Li X, Klaus H. The energy and water nexus in Chinese electricity production: A hybrid life cycle analysis. Renewable and Sustainable Energy Reviews. 2014;**39**:342-355. DOI: 10.1016/j.rser.2014.07.080

[33] Turconi R, Boldrin A, Astrup TF. Life cycle assessment (LCA) of electricity generation technologies: Overview, comparability and limitations. Renewable and Sustainable Energy Reviews. 2013;**28**:555-565. DOI: 10.1016/j.rser.2013.08.013

[34] Littlefield J, Cooney G, Marriott J. Power Generation Technology Comparison from a Life Cycle Perspective (No. 2012-1567). USA: Energy Sector Planning and Analysis (ESPA)/National Energy Technology Laboratory (NETL); 2013. DOI: 10.2172/1515245

[35] Gagnon L, Bélanger C, Uchiyama Y. Life-cycle assessment of electricity generation options: The status of research in year 2001. Energy Policy. 2002;**30**:1267-1278. DOI: 10.1016/S0301-4215(02)00088-5

[36] Hondo H. Life cycle GHG emission analysis of power generation systems: Japanese case. Elsevier - Energy. 2005;**30**:2042-2056. DOI: 10.1016/j.energy.2004.07.020

[37] Raadal HL, Gagnon L, Modahl IS, Hanssen OJ. Life cycle greenhouse gas (GHG) emissions from the generation of wind and hydro power. Renewable and Sustainable Energy Reviews. 2011;**15**:3417-3422. DOI: 10.1016/j.rser.2011.05.001

[38] Like W, Yuan W, Huibin D, Jian Z, Rita YML, Zhihua Z, et al. A comparative life-cycle assessment of hydro-, nuclear and wind power: A China study. Applied Energy. 2019;**249**:37-45. DOI: 10.1016/j.apenergy.2019.04.099 ISSN 0306-2619

[39] Geller MTB, Meneses AAS. Life cycle assessment of a small hydropower Plant

in the Brazilian Amazon. Journal of Sustain Development of Energy, Water Environment System. 2016;**4**: 379-391. DOI: 10.13044/j.sdewes. 2016.04.0029

[40] Varun RP, Bhat IK. Life cycle energy and GHG analysis of hydroelectric power development in India. International Journal of Green Energy. 2010;7(4):361-375. DOI: 10.1080/ 15435075.2010.493803

[41] Dones R, Baues C, Bolliger R, Burger B, Heck T. Life Cycle Inventories of Energy Systems: Results for Current Systems in Switzerland and Other UCTE Countries (No. 5), Ecoinvent Report. Dübendorf - CH: Paul Sherrer Institute Villigen, Swiss Centre of Life Cycle Inventories; 2007. Available from: http:// ecolo.org/documents/documents_ in_english/Life-cycle-analysis-PSI-05. pdf. [Accessed: August 2, 2018]

[42] Pascale A, Urmee T, Moore A. Life cycle assessment of a community hydroelectric power system in rural Thailand. Renewable Energy. 2011;**36**:2799-2808. DOI: 10.1016/ j.renene.2011.04.023

[43] Gallagher J, Styles D, McNabola A, Williams AP. Current and future environmental balance of small-scale Run-Of-River hydropower. Environmental Science & Technology. 2015;**49**:6344-6351. DOI: 10.1021/acs. est.5b00716

[44] Ribeiro FM, Silva AS. Life-cycle inventory for hydroelectric generation: A Brazilian case study. Journal of Cleaner Production. 2010;**18**:44-54. DOI: 10.1016/j.jclepro.2009.09.006

[45] Pant R, Aggarwal S, Joshi K. Life cycle assessment of small hydro power plants in Uttarakhand. International Journal of Current Engineering Technology. 2016;**6**:289-294. Available from: inpressco.com/life-cycle-assessment-of-small-hydro-power-plants-in-uttarakhand/. [Accessed: September 1, 2018]

[46] Kadiyala A, Kommalapati R, Huque Z. Evaluation of the life cycle greenhouse gas emissions from hydroelectricity generation systems. Sustainability. 2016;**8**:539. DOI: 10.3390/ su8060539

[47] Kumar A, Schei T, Ahenkorah A, Caceres Rodriguez R, Devernay JM, Freitas M, Hall D, et al. Hydropower, in: IPCC Special Report on Renewable Energy Sources and Climate Change Mitigation. NY, USA: Cambridge University Press; 2011. Available from: www.ipcc.ch/ report/renewable-energy-sources-and-climate-change-mitigation/. [Accessed: September 1, 2018]

[48] Zhou J. Life Cycle Assessment of Greenhouse Gas Emissions from Nam Theun 2 Hydroelectric Project in Central Laos (Masters Project). USA: Duke University; 2011. Available from: www. hdl.handle.net/10161/3595. [Accessed: September 5, 2018]

[49] Steinhurst W, Knight P, Schultz M. Hydropower Greenhouse Gas Emissions-State of the Research. EUA: Synapse Energy Economics, Inc.; 2012. Available from: www.nrc.gov/ docs/ML1209/ ML12090A850.pdf. [Accessed: September 5, 2018]

[50] Galy-Lacaux C, Delmas R, Kouadio G, Richard S, Gosse P. Long-term greenhouse gas emissions from hydroelectric reservoirs in tropical forest regions. Global Biogeochemical Cycles. 1999;**13**:503-517. DOI: 10.1029/1998GB900015

[51] Tundisi JG, Santos MA, Meneses CFS. Tucuruí reservoir and

hydroelectric power plant. Sharing experiences and lessons learned in Lake Basin management. Management Experience Lessons Learned Brief. 2003;**1**:1-20. Available from: <https://www.researchgate.net/publication/283091271_Tucurui_Reservoir_and_Hydroelectric_Power_Plant_Management_Experience_Brief>. [Accessed: September 5, 2018]

[52] Fearnside PM. Hidrelétricas Como "Fábricas de Metano" e o Papel dos Reservatórios em Áreas de Floresta tropical na Emissão de gases de Efeito Estufa. Oecologia Bras. 2008;**12**: 100-115. Available from: www.philip.inpa.gov.br/publ_livres/2008/Hidreletricas%20fabricas%20de%20metano.pdf. [Accessed: August 19, 2018]

[53] Fearnside P M. Hidrelétricas na Amazônia - Impactos Ambientais e Sociais na Tomada de decisões Sobre Grandes Obras. Manaus: INPA; 2015. Available from: www.philip.inpa.gov.br/publ_livres/2015/Livro-Hidro-V1/Livro%20Hidrel%C3%A9tricas%20V.1.pdf. [Accessed: October 19, 2018]

[54] Faria F, Jaramillo P, Sawakuchi HO, Richeu JE, Barros N. Estimating greenhouse gas emissions from future Amazonian hydroelectric reservoirs. Environmental Research Letters. 2015;**10**:124019

[55] Finkbeiner M. Product environmental footprint-breakthrough or breakdown for policy implementation of life cycle assessment? International Journal of Life Cycle Assessment. 2014;**19**:266-271. DOI: 10.1007/s11367-013-0678-x

[56] Zhang Q, Karney B, MacLean H, Feng J. Life-cycle inventory of energy use and greenhouse gas emissions for two hydropower projects in China. Journal of Infrastructure Systems. 2007;**13**:271-279. DOI: 10.1061/(ASCE)1076-0342(2007)13:4(271)

[57] Dorber M, May R, Verones F. Modeling net land occupation of hydropower reservoirs in Norway for use in life cycle assessment. Environmental Science & Technology. 2018;**52**(4):2375-2384. DOI: 10.1021/acs.est.7b05125

[58] Fthenakis V, Kim HC. Land use and electricity generation: A life-cycle analysis. Renewable and Sustainable Energy Reviews. 2009;**13**:1465-1474. DOI: 10.1016/j.rser.2008.09.017

[59] Gagnon L, Vate JVF. Greenhouse gas emissions from hydropower. Energy Policy. 1997;**25**:7-13. DOI: 10.1016/S0301-4215(96)00125-5

[60] Fonseca IF. A Construção de Grandes Barragens no Brasil, na China e na Índia: Similitudes e Peculiaridades dos Processos de Licenciamento Ambiental em Países Emergentes. Rio de Janeiro: IPEA: IPEA. Instituto de Pesquisa Econômica Aplicada; 2013. p. 22 ISSN 1415-4765

Chapter 6

Potential of Nonlinear Dynamics Tools in the Real-Time Monitoring of Large Dams: The Case of High Enguri Arc Dam

Tamaz Chelidze, Teimuraz Matcharashvili, Ekaterine Mepharidze, Levan Mebonia, Mirian Kalabegashvili and Nadezhda Dovgal

Abstract

Large dams are grand structures with a complex nonlinear dynamic behavior. These nonlinear effects, though relatively small, can be very important for the analysis of dam mesoscopic damage accumulation using available monitoring data, namely, the time series of strains/tilts of the dam structure in response to the periodic filling/draining process of the reservoir. The authors derive the characteristics of the unknown dynamics using the time series of tilts and strains of a structure by means of recurrence plots (RPs), recurrence quantification analysis (RQA), Lempel-Ziv complexity (LZC), mutual information (MI), detrended fluctuation analysis (DFA), and singular spectrum analysis (SSA) for studying dam dynamics. Anomalies in the nonlinear dynamics characteristics of the measured time series of the tilts/strains of dam during the reservoir regular filling/discharge process may signal the abnormal behavior of the object. These methods were used for the analysis of the monitoring data of the 271-m-high Enguri arch dam, still one of the highest (in its class) dams in the world, which was built in the canyon of Enguri river (West Georgia) in the 1970s. Since 1996, the European Centre "Geodynamical Hazards of High Dams" of the Council of Europe has been operating on Enguri dam. Since 2020, the International Project DAMAST: Dams and Seismicity has been going on in the Enguri dam area.

Keywords: real time dam tilts/displacements, nonlinear dynamics, diagnostic tools

1. Introduction

The high Enguri arch dam (271 m) was erected in the 1970s in the canyon of the river Enguri. The dam area is a zone of high seismicity with the moment magnitude $M_w = 8$. It is close (several hundred meters) to the large Ingirishi active fault (**Figure 1**).

The high seismic and geodynamical activities together with the dense population downstream of the dam made the Enguri dam with its one billion cubic meters

IntechOpen

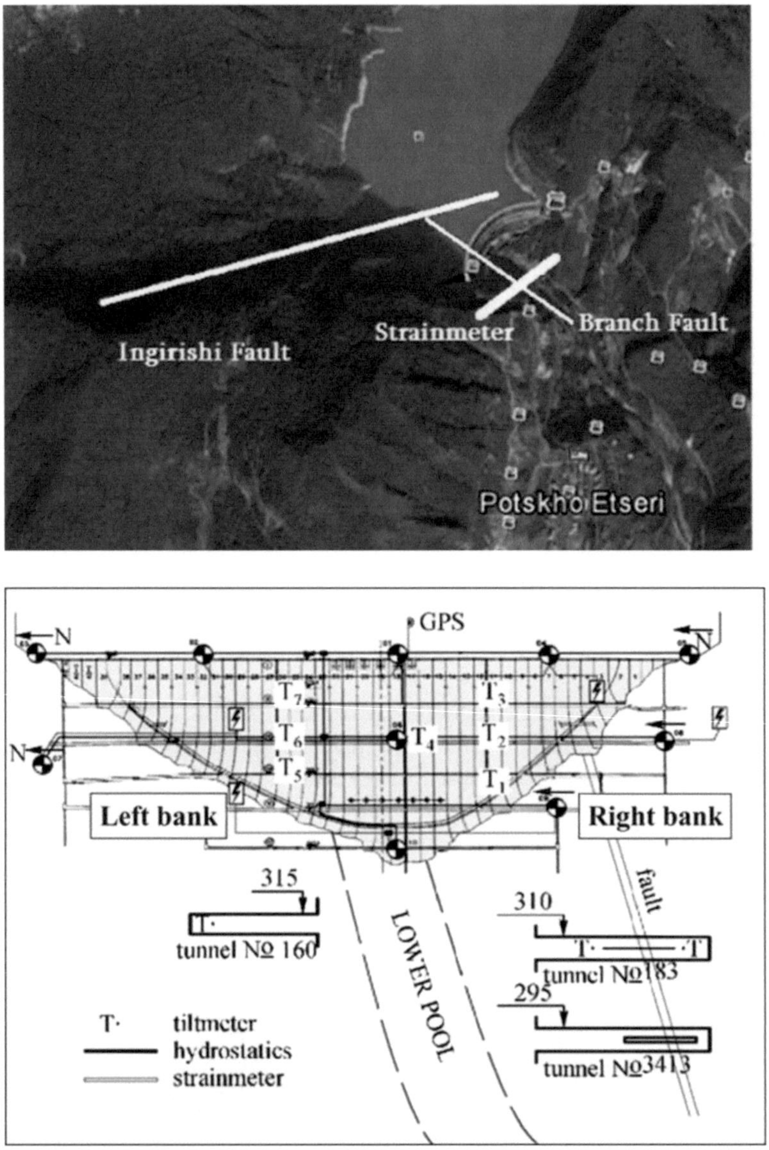

Figure 1.
Space image of Enguri dam area (upper) and the cross-section of Enguri dam with a monitoring system, a view from the upper pool (lower): T—Tiltmeters; S—Laser strainmeter.

water reservoir a potential source of a major catastrophe in Georgia. In turn, this means that the dam should be under permanent monitoring. The first geological/geophysical prospecting of the Enguri Dam area began in 1956 by the Moscow Institute "Hydroproekt" and its branch in Tbilisi. Since 1970, the M. Nodia Institute of Geophysics (MNIG) has been actively participating in the solution of problems related to dam planning and monitoring.

In 1996, the European Centre "Geodynamical Hazards of High Dams" (GHHD) with the International Test Area (EDITA) was organized in Georgia by the Council of Europe (www.coe.int/t/dg4/majorhazards/centres). Since 2020, Georgia has been participating in the International Project DAMAST: Dams and

Seismicity—Technologies for safe and efficient management of hydropower reservoirs (http://ig-geophysics.ge/IG/wp-content/uploads/2017/08/). Partners plan to investigate the impact of tectonic, seismic, and technological processes on the Enguri dam and its environment in order to assess the possible sources of hazard to the structure and elaborate the modern safety concepts of large engineering constructions in the regions of high seismicity. The project provides for the implementation of the innovative methods: satellite-based global navigation satellite system (GNSS) and interferometric synthetic aperture radar (InSAR) systems, microseismic observations for monitoring tectonic, and man-made deformations of the Enguri dam and surrounding area as well as detail study of the reservoir induced microseismicity.

For the practical use of the monitoring data, we designed the program DAMWATCH [1] for highlighting the anomalies in the dynamic characteristics of the measured time series of the tilts/strains of dam during the reservoir regular filling/discharge process, which may signal the abnormal behavior of the object. To detect false anomalies caused by the regular technological (repair, planned water discharge, etc.) and weather factors (waves generated in the reservoir by seiches) and to separate them from the signals of possible dangerous processes in the dam and its foundation (anomalous local deformations, etc.), it is necessary to keep the detailed chronological list of technological and weather impacts.

Finally, the authors would like to note that the Enguri dam area is an amazing natural laboratory, where one can investigate in detail the dam response to both the natural (tectonic, meteorological) and geotechnical (lake load-unload) impacts, i.e., the structural, geodynamical, and seismic reaction to a controllable loading of the Earth crust.

2. Monitoring system, equipment, methods

The creation of an effective monitoring/early warning system (EWS) is a necessary component for providing a secure exploitation of the large dams [2, 3]. The M. Nodia Institute of Geophysics (MNIG) and Georgian-European Centre GHHD, operating in the frame of Open Partial Agreement on Major Disasters at the Council of Europe (EUR-OPA) developed the permanent multi-disciplinary geodynamical-geophysical telemetric monitoring network (**Figure 2**). This cost-effective early warning system (EWS) consists of sensors (tiltmeters, APPLIED GEOMECHANICS Model 701-2, laser strainmeters), terminal and central controllers and GSM/GPRS Modem—for sending information to the diagnostic center (**Figure 2**).

The Enguri dam area is located in the continent–continent collision zone of the eastern segment of the Mediterranean Alpine Belt, between the converging Africa-Arabian and Eurasian lithosphere plates [3]. The branch fault (**Figure 1a, b**) of the main east-northeast (ENE), west-southwest (WSW) oriented Ingirishi fault crosses the foundation of Enguri dam and threatens the dam safety seriously. In order to monitor permanently the fault behavior, a quartz strain-meter crossing the fault zone (FZ) was installed 100 m downstream of the dam foundation in December 1974. The fixed and free parts of the strainmeter are located on the opposite sides of the FZ in the intact massive; it records fault zone extension (FZE). After 2000 the free end of the tube is equipped with the laser system (Laser model R-39568, Green HeNe Laser, 633 nm and Laser Position Sensor OBP-A-9 L) instead of old-capacity sensors.

The displacements' sensitivity of the strainmeter system is of the order of 1 μm/mm. The monitoring system of the Enguri dam and its foundation includes also, besides tiltmeters and strainmeters, a network of piezometers and reverse plumblines

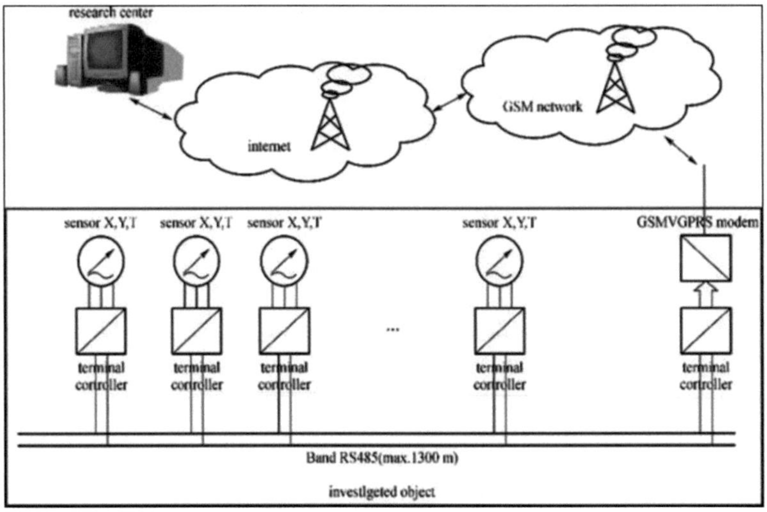

Figure 2.
The monitoring system of Enguri dam consists of sensors (tiltmeters, APPLIED GEOMECHANICS model 701-2) and strainmeters, which are connected to terminals and central controllers. GSM/GPRS modem transmits the data to the diagnostic center.

[4], weather station, and water-level (WL) gauges for monitoring water level in the lake.

Engineering studies of the dam and its foundation. To study the current state of the rock mass in the right bank of the dam foundation, several wells have been drilled, and engineering-geological and geophysical analyses of cores were carried out in 2019. After filling the reservoir and the emergence of a new aquifer under the dam the pressure variation at drawdowns and fillings of the lake cause not only the compaction/extension of the massive, but also the partial dissolution of the crack calcite filler in the mass. According to the results of wells' video-logging in the right bank, it should be noted that the rocks in the upper part of the wells look, for the most part, very strongly destroyed and cavernous. Probably, during dam operation, many calcite veinlets were eroded, and their filler turned into a powder and partially dissolved: as a result, a large number of small voids appeared, which are visually traced on the surface of the well. The presence of the right-bank fault led to the superposition of the influence of two natural factors (fault, veinlets of quartz in limestones) and a series of artificial ones (the load-unload cycles, accompanied by the filtration of water) that arose due to the impact of the system operation. According to the completed documentation of the drilled wells, the supposed most vulnerable near-contact part of the dam base is in a satisfactory condition. The results of ultrasonic logging confirm the indications of video logging and geological documentation data.

To assess the deformation processes of the foundation, a numerical analysis was carried out using a three-dimensional model of the "dam-foundation" system, taking into account operational loads and the relative degradation of the deformation and elasticity modulus in the "load-unload" cycles. The considered factors affect the rotation of the deformation vector of the left dam abutment toward the riverbed and the right-bank abutment—to the depths of the massif.

The residual displacements of the dam and foundation during filling and drawdown of the reservoir after 40 years of operation exceed the displacements obtained by the "dam-foundation" system during the first filling of the reservoir to the design level. At the same time, the graphs of the calculated displacements of the characteristic points on the crest of the dam indicate the damping nature of the growth of deformations last years, approximately from 2011.

Thus, the features of the canyon shape, the geological properties of the foundation, the geometry of the arch dam and deformations accumulated during many years of dam operation contribute to the formation of an inhomogeneous stress field in the rock mass of the left-bank abutment and a less favorable development of deformation processes compared to the right bank.

3. Results. Monitoring data

3.1 Dam stability measures

Traditional approach. The standard dam diagnostics based on the linear elasticity model as a rule ignore the microscopic inhomogeneity of the dam material. Elastic properties of such inhomogeneous materials at small concentration of defects are described by the theory of effective media [5] and at the high density of defects - by percolation elasticity theory [6]. The standard static analysis of monitoring data was performed using a 3D finite-element model (FEM) for the dam-foundation system. The correlation between natural observations and theoretical calculation at lower levels of the dam is acceptable, but there are significant deviations from the calculated values in the upper dam levels (**Table 1**). The displacement observations' data obtained by two different methods (tiltmeters and plumblines) are in satisfactory agreement (**Table 1**). The diagnostics of dam safety according to [7] considers three stages: 1—the normal state (N); 2—the strain is maximal allowable (MAS); 3—the strain corresponds to a pre-failure stage (PFS). The most grave diagnose PFS mean that after repeated investigation the exploitation regime should be changed up to the 0full shutoff.

The analysis of the table leads to the following conclusions: a. at the WL 360 m, all the displacements are less than critical values (class N); b. at the level 402 m, only in the central 18-th section, the displacements are close to critical ones (class MAS); 3. at the highest level (475 m), displacements of the side sections are larger than critical by 8–13 mm, i.e., at this level, the state should be considered as class MAS or "faulty." At the same time, the dam performs normally, and there are no signs of significant damage, which mean that the theoretical model should be updated.

The most important drawback of this approach is that the static design model does not predict nonlinear effects, i.e., the existence of hysteresis phenomenon and the possible impact of dam material aging.

3.2 Long- and short-term diagnostic tools

Long-term diagnostic tools. In the standard approach to dam safety assessment the response of real strain/tilt data of construction is compared to the design values based on Hooke's linear-elastic constitutive model. If the stress–strain response during load-unload cycle follows the hysteretic cycle that signals on the significant deviations of observed data from the static elasticity model (**Figure 3**, **Table 1**). The generally accepted explanation of hysteresis is thermo-elasticity, though the mesoelasticity theory (see below) can be more appropriate [8].

Other peculiarities of the load-unload process (time lag between water load and strains in the dam, details of strain build up and relaxation)can also be used for long-term diagnostics.

Dam Level Observation Level, m	Section 12			Section 18			Section 26		
	Observed plumbline data, mm	Observed tiltmeter Data, mm, sec	CAV mm, sec	Observed plumbline data, mm	Observed tiltmeter Data, mm,sec	CAV, mm	Observed plumbline data, mm	Observed tiltmeter Data, mm. sec	CAV, mm
360	20	11 (38 ± 5.1)"	89 (122)"	35			15	14 (46 ± 5.6)"	88
402	40	32 (63 ± 4.5)"	59 (112)"	60	55 (70 ± 3.9)"	55	30	37 (74 ± 4.1)"	58
475	60	48 (56 ± 8.7)"	31 (182)"	65			55	42 (55 ± 5.5)"	26

Table 1.
Comparison of observed plumbline horizontal displacements and tiltmeters data at the maximal water level (WL) in the lake (510 m) for three sections of Enguri dam with theoretical critical admissible values (CAV) of plumblines (2008).

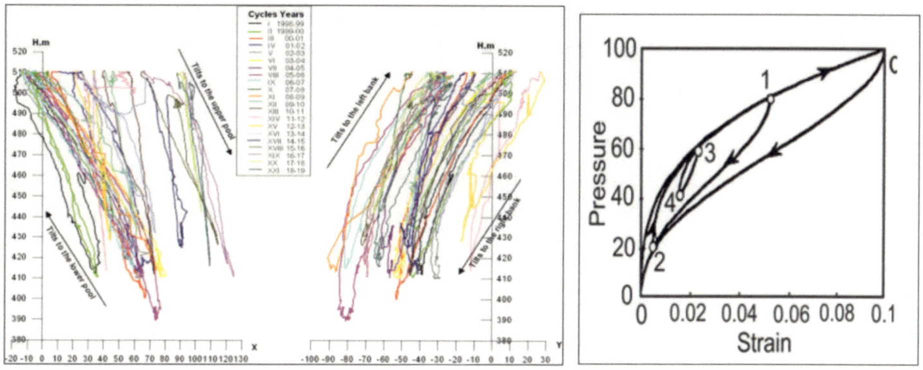

Figure 3.
Left—Horizontal axis—tilts, registered in the section 12, mark 402 of Enguri arch dam (in sec), along (X) and normal (Y) to the dam crest; vertical axis—water level in the lake H in meters, during 21 seasonal cycles (1998–2019); Right—the history of strain during the pressure cycle applied to a system, containing many hysterons. Note the similarity of the real and theoretical hysteresis curve structures.

Short-middle-term diagnostic tools. Besides above indicators, there are some short-term features in the low-frequency (LF) part of the spectrum of natural frequencies of dam vibration (dam tremble), which reflect the response of dam structure to various impacts such as water discharge, wind and waves, turbine rotation or ambient seismic noise. This can be used for the operative/express dam state diagnostics, namely, LF dam vibration spectrum (amplitudes) should respond to the change of the state of dam damage (some frequencies disappear; new harmonics appear, etc.).

3.3 Nonlinear Response—Hysteresis

Figure 3 presents the tilts, registered in the body of Enguri Dam, (Section 12, mark 402) in two directions, along (X) and normal (Y) to the dam crest versus water level (WL) in the lake H in meters, during 20 seasonal cycles (1998–2018). The theoretical interpretation of experimental hysteretic tilt-stress dependences at present is mainly taking into account thermal effects, though, in our opinion, it can also be accomplished by the theory of the mesoelasticity (mesoscopic nonlinear elasticity) model. In the thermal theory of hysteresis, the dam material is considered as homogeneous media. In reality, the dam material is heterogeneous and contains an enormous number (10^9–10^{12}) of the so-called mesoscopic structural features (mainly compliant microcracks) in a square centimeters, which means that macroscopic elastic properties of material depend strongly on the behavior of microcracks under varying stress. Such structures as a rule manifest nonlinear hysteretic elastic behavior, which is connected with a specific response of the microcrack system to stress variation, namely, asymmetric response to load and unload (**Figure 3** right). The formal approximation of experimental annual hysteresis data can be accomplished using the Preisach-Mayergoyz (P-M) phenomenological model [8]. In the P-M model, the system is represented by the complex of hysteretic elastic units or hysterons.

The hysteresis in the stress–strain dependence stems from difference in the number of open/closed defects (hysterons) at increasing and decreasing pressure. Like in all hysteretic phenomena, this difference in load-unload response is due to irreversible energy losses during the process of crack growth (or generation) so that for restoring the population of closed cracks at unloading it is necessary to apply an additional pressure ΔP. As the value of ΔP depends on the concentration of hysterons, the opening of hysteresis loop reflects the level of material damage. One can use this parameter for diagnostics purposes [8]. The above P-M model corresponds to the reversible hysteresis process (**Figure 3**, right), which means that after the loading/unloading cycle, the system returns to the initial state. In the real systems, the process is not reversible as besides elastic strain, there is also residual deformation. Consequently, in repeating load-unload cycles the positions of hysteresis loops initial and final points shift along the strain axis, as it is evident in **Figure 3**. The shift of hysteresis loops can be used as a measure of residual strain (damage rate) in a given material.

The variation of the slip rate during load/unload of dam lake can also be related to the change of the state of fault filling material. During the operation of the dam, under the influence of loads and physical-chemical leaching processes, a partial transformation of the vein calcite into a powdered state occurred, which can increase strain rate (SR). On the other hand, the penetration of hydrothermal solute can lead to the formation of numerous solid veins and veinlets of quartz and calcite, increasing the fault stability (increasing friction of fault faces). Some of the observed secondary short-term hysteresis loops (**Figure 3**) occur too fast, which make the relatively slow physical–chemical mechanism less probable.

The decisive validity test of two models—thermoelasticity and hysteron (mesoscopic nonlinear elasticity)—would be finding hysteretic behavior at the almost constant temperature conditions: that should be connected mainly with the mesoelasticity effect due to the fast variation of WL in the reservoir, though we cannot exclude the contribution of thermoelastic component also during the long load-unload cycle.

3.4 Dam reactions—Monitoring data

Reactions of the dam structure to the natural and technogenic strains. As was mentioned, the generally accepted approach to the dam safety problem is to compare the observed strain, characteristic frequencies, etc., to the calculated critical values. The calculated dominant frequency (main mode) on the crest of the Enguri dam is around 2 Hz, which is in a good accordance with the results of analysis of accelerograms recorded during the last 2005 Racha earthquake (EQ). At the same time, our data obtained by 701 platform tiltmeters at the high sampling rate (1 per minute) show that EQs excite vibrations at the much lower frequencies also.

Technical (exploitation) impacts. The tiltmeter recordings show that some technical operations, including trivial energy cuts as well as scheduled water discharges for reducing WL at the intensive inflow may induce significant effects on the tilts/strains of the dam complex [4].

Earthquakes. The analysis of seismic record of the October 6, 2005 Racha event M = 6, occurred at the distance 100 km, shows that two mean frequencies can be identified ~1.6–1.7 and ~2.1–2.2 Hz, which mean that for relatively high frequencies, the dam response is elastic, unlike its hysteretic response to slow strains. At the same time, our data obtained by 701 platform tiltmeters at a high sampling rate (1 per minute) show that EQs excite dam vibrations at the much lower frequencies also. The analysis of low-frequency (LF) response spectra indicates that the network consisting of such tiltmeters extends significantly the frequency range of nonbroadband seismic devices from 100 Hz to a quasi-static range; this can give enormous additional information on the dam state.

The distance from the great Tohoku EQ epicenter to Enguri dam is 7800 km. The Enguri dam responds to the event by appreciable vibrations (**Figure 4**, right). Only

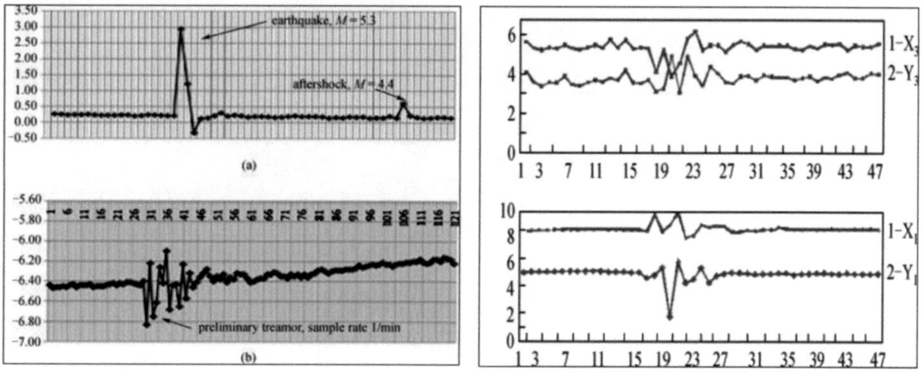

Figure 4.
Left, (a)—expanded recording of Enguri dam tilts due to the local Vani EQ: Main EQ—Vani, M = 5.3, 19.01.2011 (distance to dam 80 km) and aftershock M = 4.4; (b)—preliminary tremor 13 hours before main Vani event; sample rate - 1/minute. Right—Dam tilts at Tohoku (Japan) EQ, M = 8.9 march 11, 2011, at 05:46:23 UTC on the stations X_1 Y_1 and X_2 Y_2; sample rate 1/10 min.

the trace of the long period seismic wave components (30 sec and more) was registered due to a low sampling rate.

Seisches. In the enclosed body of water, the disturbances of water body by wind, earthquakes, etc., generate a standing wave or seiches, which, in turn, excite the vibrations of the dam. The seiche-generated vibrations on the Enguri dam were recognized by their co-incidence with strong winds. The disturbance from seiches can be interpreted in the following way: the longest natural period T of seiches in an enclosed rectangular body of water is $T = 2L / \sqrt{gh}$, where L is the length, h is the average depth of the body of water, and g is the acceleration due to gravity. Inserting corresponding values $L = 25$ km and $h = 50$ m for the Enguri lake, we get for the period of seiches $T = 38$ min, close to the observed values: 30–40 min (**Figure 5**).

Intensive water discharge. The discharge of excess water through a slide gate (dam outlet) also causes tremors in dam body: their signature is the maximal intensity in the middle level of dam and smaller intensity at both the upper and lower levels.

Fault strain change due to intensive precipitation. The strainmeter installed on the fault crossing the foundation of the dam registered abrupt strong deviations of the FZ strain rate from the regular background values during strong precipitation events. The observed strain rate exceeds the regular pattern connected with the lake load-unload by several orders of magnitude. The anomaly could cause serious alarm, if one does not know the nature of this transient [9].

A detailed study of long- and short-term anomalies is important, as they can be precursors of a serious trouble. At the same time, some anomalies are of trivial technical origin connected with dam exploitation, as it was shown above. Thus, it makes sense to keep a systematic record of anomalies and compare it to usual technical "noise", as well as to the impacts of meteorological factors (strong winds, intensive precipitation, etc.) in order to single out the really dangerous precursors of construction damage.

3.5 Strains in the dam foundation

Figure 6 presents the strain (extension) history of the fault zone (FZE) under the dam during almost half-century, which demonstrates that dramatic changes can happen in its dynamics with time. The record of the fault extension rate shows that significant changes—namely, acceleration and deceleration—occur in 1984, 1987,

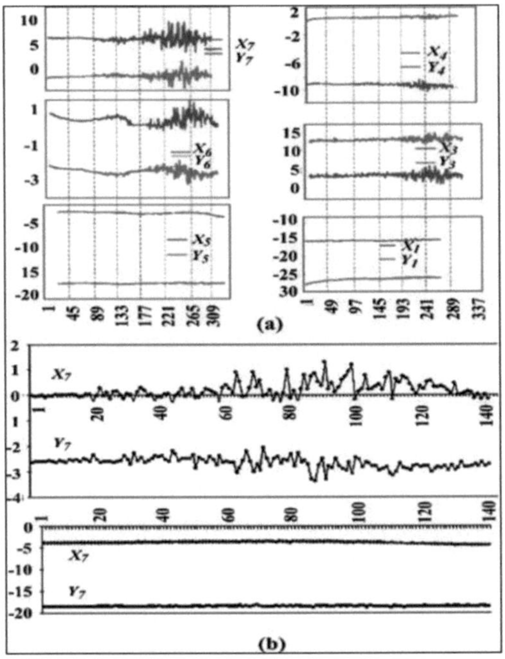

Figure 5.
(a) Example of perturbations in the dam tilts due to strong wind (seiches): Readings from 1 to 171 correspond to a calm weather and readings from 171 to 292—To a stormy one. The sample rate—1/10 min.); (b) expanded record of tremor caused by lake water perturbations due to seiches—The mean period of perturbations is of order of 30–40 min.

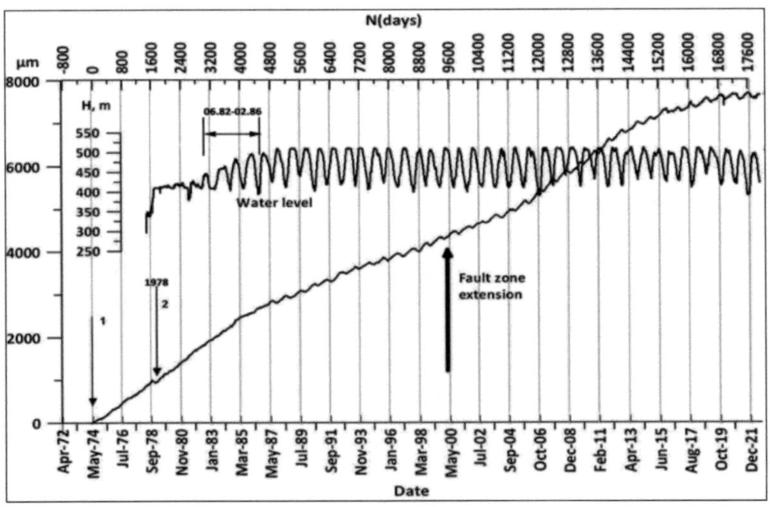

Figure 6.
WL in the Enguri lake (H, meters) from 1978 to 2021 (upper curve) and the data on the extension/compaction (in micrometers) of the branch of Ingirishi fault, crossing the foundation of the dam, from 1974 to 2021 (lower curve). Arrow 1 corresponds to the start of strainmeter monitoring.

2004, 2013, and 2017. There are two main components of the strain rate curve—the piecewise-linear changing increment of the fault opening and superimposed on its quasi-periodic strain variation.

N	Periods, months–years	Number of days within the periods 1–7; in the brackets the same from the zero day (01.04.1964) to the end of a given period	Strain rate microns/year
1	May 1974– Sep 1984	1600 (3807)	250
2	Sep 1984–May 1987	973 (4780)	160
3	May 1987–Sep 2004	6332 (11,112)	147
4	Sep 2004–Apr 2013	3379 (14,400)	233
5	Apr 2013–June 2017	1551 (16,000)	147
6	June 2017-Dec 2021	1160 (17,600)	44

Table 2.
Periods with a different strain rates on the fault.

Table 2 presents the time history of the fault zone extension (FZE) linear component, (defined as a straight line approximating the general trend) beginning from 1974, i.e. before the erection of the dam to 2021.

We discuss the possible explanations of the strain rate change we discuss later on.

Intermediate-term strains of the fault zone. **Figure 6** shows the pattern of dependence of fault zone extension (FZE) on the water level *H*. We used the detrended from the linear part FZE trajectory (i.e., the quasi-periodic decoration of the FZE) during one year's recharge period (30 Apr 2013-30 Sep 2013), which is more or less repeated from year to year to assess the reaction of fault strain to Δ WL variation. According to the data, WL rise in the lake by 80 m leads to the (almost reversible) change in an FZC equal to 70 μm. Accordingly, the fault sensitivity to WL variation is FZE/ Δ WL $\approx 0.875 \mu m$ /m. Note that, actually, the FZ displacement plot is hysteretic and the above approximation describes only the close to linear (longest) section of the hysteresis loop (excluding its short initial and final nonlinear sections).

4. Discussion

4.1 Interpretation of monitoring data

To ensure the operative statistical and dynamical investigation of dam stability problem, the modern methods of nonlinear dynamics analysis of strain/tilt time series are appropriate to use [10, 11]. Contemporary methods of time-series analysis include detrended fluctuation analysis (DFA) [12], Recurrence Plots (RP) and Recurrence Quantitative Analysis (RQA) [13], algorithmic (Lempel-Ziv) complexity measure (LZC) [14, 15], mutual information (MI) [16, 17], singular spectrum analysis (SSA) [18], and Mahalanobis distance (MD) [19].

The suggested methods for dam safety analysis are informative because: i. the strains/tilts and other geotechnical parameters of the dam as well as dynamic characteristics of strains/seismic activity in the dam area manifest as a rule quasi-periodic variations in time due to seasonal cycles; ii. due to quasi-periodic dynamics of water load, the related strain/tilt time series should follow relatively stable orbits in the phase space; and iii. the strong decrease of time-series determinism or significant deviation from the stable orbits compared to long-term stable pattern of strains/tilts/seismic activity can signal on the beginning of instability. In the following, we present short descriptions of the used methods.

RP is based on the analysis of recurrences of the system to the certain state (or the state space location), reflected by appearance of diagonally and vertically oriented lines in the recurrence plot. This is a fundamental property of any dynamical system with quantifiable extent of determinism.

RQA enables to quantify the features of the diagonally and vertically oriented lines in the recurrence plot by means of several measures of their complexity. In this research, we calculated: i. determinism %DET—the fraction of recurrence points that form diagonal lines of recurrence plots (they show changes in the extent of determinism in the analyzed data sets); ii. laminarity %LAM, the number of recurrent points that form vertical lines: in other words, it represents the frequency of laminar states in the system; and iii. trapping time (TT)—the length and amount of the vertical structures in RP, which show how long the state of the system is trapped.

LZC calculation is used for assessing the extent of order in different data sets by transformation of a given data set to a new sequence of symbols. For this, the original data are compared to a certain threshold value (say, the median of the observed data set) and converted into a 0, 1 series. The obtained symbolic sequence is parsed to obtain distinct words, which are encoded by a specific rule.

MI—expressed in bits—measures the mutual dependence between two variables. High mutual information corresponds to the reduction of randomness—if the two data sets are independent, the MI is low. For fully independent variables, MI is zero, and for highly correlated variables, MI is of the order of one or larger.

SSA: singular spectrum analysis allows the decomposition of time series into a sum of independent components. According to [20], using SSA one can forecast new data points based on the existing information even if the time series are short and noisy. We use SSA to divide fault deformation monitoring data into two main components: long-term and short-term oscillations.

MD: Mahalanobis distance for the multivariate data measures a distance relative to the centroid—a central point of the scattered values. The farther is the given data point from the centroid the larger is MD. Usually, the MD is used to discover outliers in multivariate data (large values of MD). This points to the anomalous behavior of the system.

ML: machine learning in the past few years has been used extensively for the prediction of the evolution of nonlinear system learning from the existing past data. ML allows analyzing the behavior of a dynamic system even if it is impossible to solve the problem in a closed analytical form. At the same time, the application of data-driven ML can lead to unveiling of the associated nonlinear dynamics regime [21, 22].

We used most of the above-mentioned nonlinear dynamics methods for the analysis of Enguri dam temporal behavior [9, 23–30].

4.2 Nonlinear dynamics of the dam structure tilts

In order to test the sensitivity of selected methods and assess the effects of external influences on the strain dynamics of the dam foundation, strain data sets in the foundation of Enguri HPP in the period from 1974 to 2009 are considered. **Figure 7** illustrates RQA determinism (RQA%DET), %LAM, and trapping times (TTs). It is evident that for the different stages of dam construction and reservoir filling, significant quantitative differences occur in the recurrence attributes of phase space structures reconstructed from the fault strain datasets. Note that nonlinear dynamics of strain time series demonstrate a decrease of initial high enough level of determinism, laminarity, and trapping time during initial filling of lake with a minimum around 1979. After this, all the three parameters increase to the maximal values around 1984 and stay high till

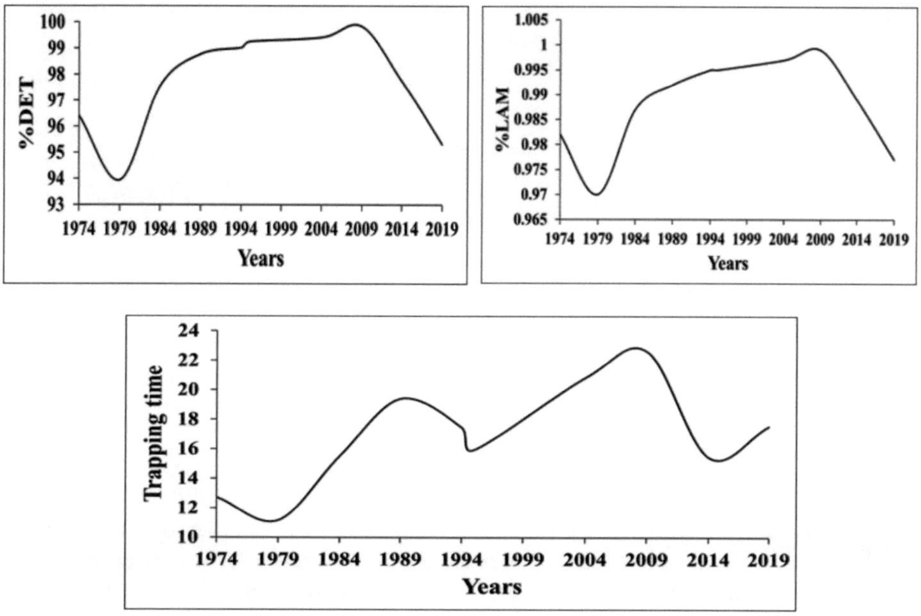

Figure 7.
RQA characteristics of the displacement of Enguri Dam foundation in 1974-2013: 1. %DET; 2. %LAM; and 3. trapping times (TT).

2012—in this period, WL variations became periodic. In the following years, the ordering decreases significantly (**Figure 7**).

4.3 Nonlinear dynamics of the fault strain

Global tectonic factor. The long-term piecewise-linear trend recorded by strainmeter (**Figures 1–6**) documents a persistent separation of fault faces, extending to 7500 μm (7.5 mm) during the observation period. The dynamics of fault reflects a joint influence of two main factors: one leads to a piecewise linear displacement (the trend component) and the other one—to quasiperiodic oscillations, decorating the main trend. The linear component of FZE rate (y) depends on the time (t) following a simple equation: $y(t) = at—b$, where the coefficient a, the slope of the linear component of the fault zone extension (FZE) differs from one period to another (**Table 2**). The first trend component (the strain rate 250 μm /m) is characteristic for 1974–1978, i.e. the period before dam construction and initial lake filling. It is natural to explain it by the action of existing natural regional tectonic stress. After this period, the fault zone extension rate (FZER) is varying significantly with time, due to the impact of some non-stationary factors (**Table 2**). Probably, the variation of the strain rate of the fault reflects the interplay between tectonic force, detail fault structure and water lake pressure. The several different epochs in the strain rate (SR) history (**Table 2**) can be explained by the interplay of some strain decelerating/accelerating factors. The 10-meter-wide fault zone is filled with strongly fractured rocks. Under stress, the extremely fractured/dense granular material exhibits mechanical properties that range from fluid to solid behavior. This complex behavior is due to the formation of the force chain networks (FCNs) (networks of force-accommodating chains) under applied stress. As a rule, such interconnected force chain networks form elastic structures, percolating through a compressed granular system. The force

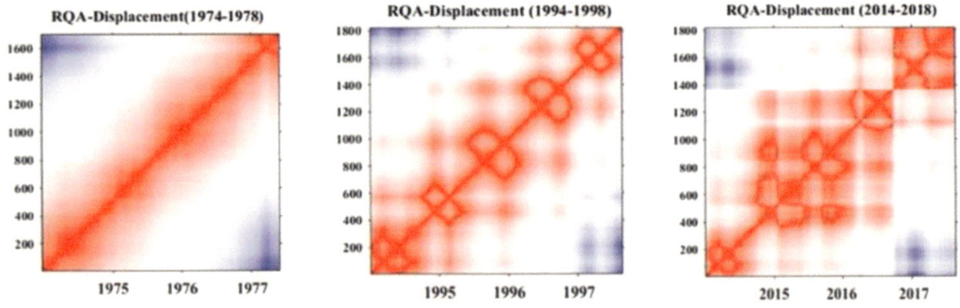

Figure 8.
Recurrence plot analysis (RP) of fault displacements for the 5-year periods: Left—before reservoir filling (1974–1978); in the next 5 years the RP is the same; middle—during regular reservoir exploitation—the identical 5 yearly cycles are present in the RPs during 1994–2013 years; right—the RP for 2014–2018 differs from previous regular patterns—The pattern is less regular.

chain network (FCN) supports the applied stress and ensures the ability of a granular system to resist deformation. Thus, the macroscopic elastic properties of such systems are determined by the development of percolating force chains. Using the above model, the first SR deceleration in 1984 can be explained by the increase in the density of percolating force chain networks (increased elasticity) of the fault gouge due to increased load on the fault at rising the water level by 150 m. The following SR acceleration 2004–2013 can be caused by the damage of the existing dense FCN, possibly due to periodically varying water load cycles leading to the so-called reversed-stress fatigue of material. We explain the presence of several different epochs in the strain rate history (**Table 2**) we explain by the interplay of the above-mentioned accelerating/decelerating factors.

Technogenic impact. Applying singular spectrum analysis (SSA) to the residual component of detrended original displacement data in the first 4500-day interval (the summary time before impoundment, during first fillings and irregular load-unload regime), we found that the leading period in the power spectrum of FZE time series is about 1.5 years [9]. After this date began the period with quasiperiodic water level variation, when the clear leading cyclic component with the period close to 1 year appears in the SSA graph: this cycle was absent in FZE dynamics before a regular variation of WL [9].

We obtain similar results using the recurrence plot analysis of the fault displacements' 45-years history (**Figure 8**): i. before reservoir filling (1974–1978) and in the next 5 years, till 1982 in the RP plots there are no clear signs of periodicity; ii. only after the beginning of the regular exploitation of reservoir, the characteristic periodic recurrent structures—the identical 5 yearly cycles—appear in the RPs, which last in 1994—2013; and iii. the RP for the last years (2014–2018) differs from previous patterns—the recurrent structures are less regular due to some technical repair works on the dam.

According to **Figure 9**, mutual dependence between WL and FZE variations increases from relatively low values—from MI=0.4 to MI=0.8—till the day mark 3090, which corresponds to the period of nonregular impoundment of the lake and to even greater values of the order of 1 or more after the start (in 1984) of the regular cyclic regime.

Note also that there is a phase lag between the maxima of WL and the maxima of FZE: the FZ displacement is retarded relative to WL maxima by several months (**Figure 10**). The delay can be explained by the fluid diffusion though the fault over the distance x from the lake to the strainmeter location (**Figure 1**). The response looks asymmetric during filling and discharge of the lake, but after subtracting the strain trend component the asymmetry disappears [9]. Using the Biot's equation of poroelasticity in the low-frequency limit [31] allows us to understand the observed time lag between fault

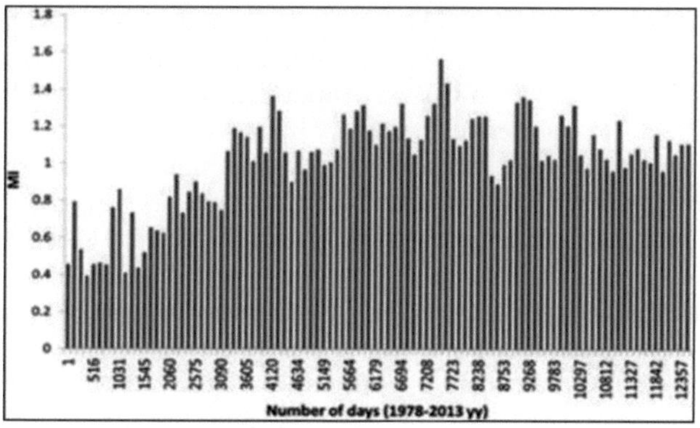

Figure 9.
Mutual information (MI) between WL and FZE variations (in bits) versus day number from the Jul-1978 to Sep-2013.

zone deformation and WL variation. According to it, the diffusion of fluid pressure perturbation P from a point source in a homogeneous hydraulically isotropic medium is described by the differential equation of fluid diffusion: $dP / dt = D\nabla^2 P$, where D is the hydraulic diffusivity coefficient. The solution of this equation is [32]: $r = \sqrt{4\pi Dt}$, where r—is the distance from the perturbation source (dam foundation) to a given point in the media and t is the time, needed for the pore pressure perturbation to reach a given point. Large phase lag between WL and FZE—200 days—points to the decisive role of poroelasticity in the fault zone dynamics [33].

The distance between the Enguri dam upper pool (pressure perturbation source) and the strainmeter site r is 100 m (**Figure 1**) and the mean value of the phase lag between the maxima of WL and the maxima of FZE during the regular impoundment regime is of the order of 200 days. Substituting these data into the above diffusion equation, we obtain for the hydraulic diffusivity D of rocks in the fault zone approximately $D \approx 5{\times}10$ m^2/s, which is close to the typical D values in similar rocks [34].

4.4 The phenomenon of reservoir-induced seismicity (RIS) synchronization

The probabilistic seismic hazard (SH) analysis of the dam area was carried out using the Cornell approach and computer program SEISRISK III [35]. For calculation

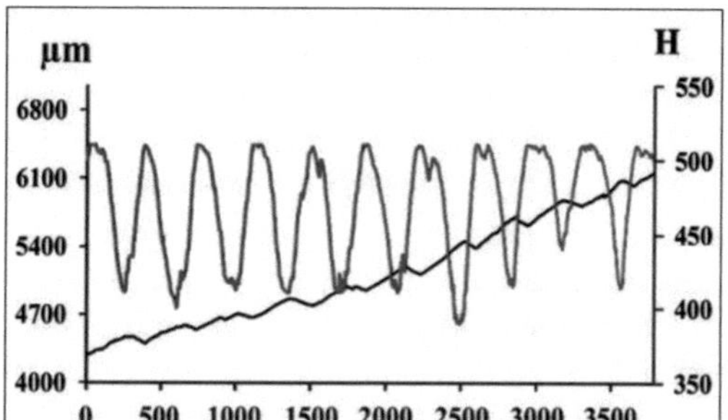

Figure 10.
WL in meters (gray) and FZE (black) variations in microns versus the number of the readout's day (the period January 1999–December 2009). Note that the maxima of the FZE are retarded relative to the maxima of water load by 200 days.

of the peak ground acceleration (PGA) attenuation model and spectral acceleration (SA) for Caucasus after was used [36]. The maps for PGA and SA for 50-year exposure time and 1%, 2%, 5%, and 10% probabilities of exceeding critical SA were compiled [37]. The SA maps were calculated for frequencies of 1.5 Hz and 2 Hz. These frequencies are close to the first modes of natural frequency of the construction, and therefore, they are the most hazardous for the dam. The above SH assessment may change after the implementation of the International Project DAMAST.

The EQs occurred close to the Enguri dam area (at the distance from 0 to 10 km) are dominated by events connected with the near zone seismicity [9]. They are linked to the reservoir water-level regime in both space and time. Almost simultaneously with the last operation of reservoir impoundment, in September 1978, an abrupt reversible compaction by approximately 90 μ m (at attaining the critical water level 100 m) was registered by the strainmeter installed on the fault, crossing the foundation of the dam.

A strong seismic activity very close to the dam: the Rechkhi EQs swarm, with a four main events of M_W 3.7-4.3, was observed in December 1979 (Δ = 0–10 km, hypocenter depth 3–10 km). It followed with a 14-month lag the fast initial recharge (September 1978) of the lake up to the critical water level (100 m) [9, 38]. According to the existing data [39, 40], such water load as a rule generates a reservoir-induced seismicity (RIS).

Reservoir-induced seismicity (RIS) may take place either due to the change of water load and corresponding change in the Earth's crust stress or by the decrease in the effective strength of the rocks by the increase in the groundwater pore pressure. These additional strains can induce an increase in both the number and the magnitude of EQs in the area close enough to the reservoir. As a rule, RIS is a transient phenomenon, which is observed after filling the reservoir with a delay, which can last from days to years. After this, the period of seismic activity decreases to the prefilling values [9] with one important exception: the time intervals between seismic events became significantly much more ordered than before periodic load/unload of the reservoir and this regime is persistent even after initial transition period [9]. This effect can be coined as Reservoir-Induced Seismicity Synchronization (RISS) [26]. The effect seems to be impossible, as the stress invoked by regular water level variation is a small external force compared to the acting tectonic forces—the main factors leading to seismic activity.

Nevertheless, the modern theory of synchronization [41] predicts that even a weak external periodic force applied to an autonomous self-sustained oscillator can adjust its rhythm. In our case, the water level in the reservoir is the source of periodic force, and the tectonic fault under dam is the autonomous self-sustained quasi-periodic system with dynamics connected with other varying factors, such as seasonal snow load, etc.

Next, the seismic process around the test area was investigated using Mahalanobis distance (MD) parameter based on the data sequences of its temporal, spatial, and energetic characteristics. The MD calculated for original data was combined with a surrogate data testing. It is useful to analyze the process based on unevenly sampled data sets used in this research, namely, sequences of increments of cumulative times of earthquake occurrences, cumulative distances, and cumulative seismic energies. After the computation of MD values, they were converted into F-values [19]. F-value is the confidence level estimate of the statistical difference between two types of data: here—between original seismic data and the set of randomized seismic catalogs around Enguri dam [23]. Then, calculated F-values were compared with the critical value for the randomized catalog, Fc. When F>Fc is calculated, then the null hypothesis that there is no separation between considered groups can be rejected, which means that there is a statistically significant difference between groups.

In **Figures 11** and **12**, the results of EQ waiting times' analysis using RQA and LZ methods are presented. There are several intervals with significant changes in the EQ waiting times' regime: 1974–1986—weak regularity in waiting times; 1986–2004—the interval of irregular seismicity; and 1998–2004—the period of increased regularity of waiting times with a maximum around 2010.

As is evident from **Figure 13**, according to MD-analysis, the seismic process becomes clearly different from a random one in the period, when the water level in the reservoir changed periodically, with a considerable delay. The ordering has become most obvious since the end of 2005. It is important that in the period prior to flooding (the end of 1977), the seismic process in the considered catalogue is indistinguishable from randomness (in all the windows, the calculated F-values are equal or below Fc). In the period of flooding and preliminary filling of the reservoir (after 1990), seismicity looks mostly different from randomness; after 2000 (i.e., several years after the beginning of the lake periodic loading), in all the windows, F is larger than Fc.

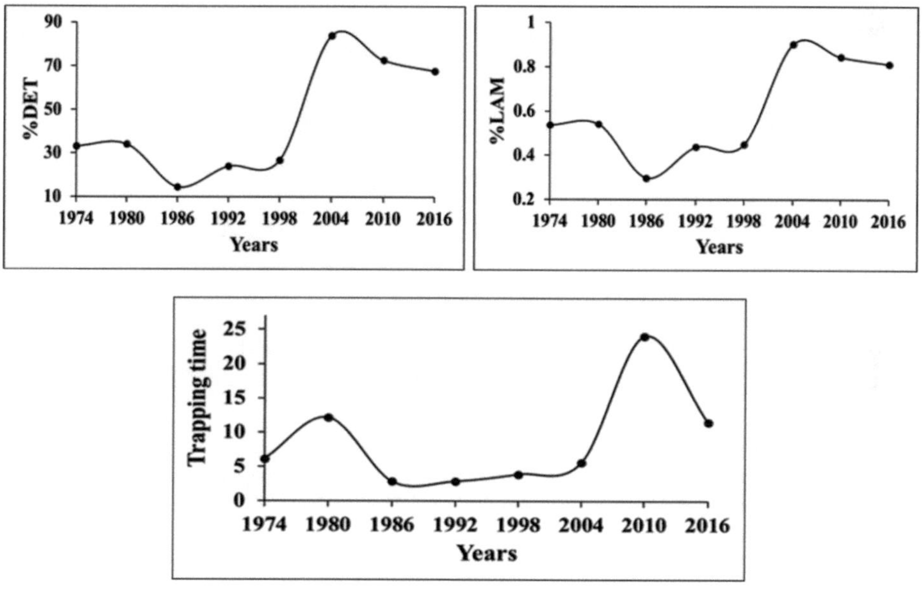

Figure 11.
RQA characteristics of earthquake waiting (interevent) times in 1974–2017 in the original seismic catalog in the Enguri dam area within 100-km radius: 1. %DET; 2. %LAM; and 3. trapping times (TT).

Figure 12.
Lempel-Ziv algorithmic complexity (LZC) analysis of waiting times between earthquakes for 1974–2017.

According to **Figures 11** and **12** the most ordered period is 1999–2014, i.e., the MD fixes the synchronization of seismicity several years later than RQA and LZ—possibly, due to the difference in methodology: the MD is based on the synchronization analysis of all the three EQ parameters (cumulative times of earthquake occurrences, distances, and seismic energies). This is distinct from the above RQA and LZ analysis, which explore only EQ waiting times' data, which seem to be synchronized with WL rhythm much easier than magnitudes and distances.

There is some connection between the EQ waiting times' regime (**Figures 11-13**) and fault zone dynamics' rate FZDR (**Figure 7**, **Table 2**). Generally, there are three main periods in both these data sets: initial (weak regularity in seismicity and constant velocity of the fault strain), intermediate (disordered seismic regime and the decoration of the linear FZDR by the irregular process of lake filling) and stable (ordered seismic regime and quasi-periodic WL variation). It seems that the ordering process of seismicity (1998–2004) lag behind the fault strain synchronization by water load (1980–1982), i.e. by approximately 20 years.

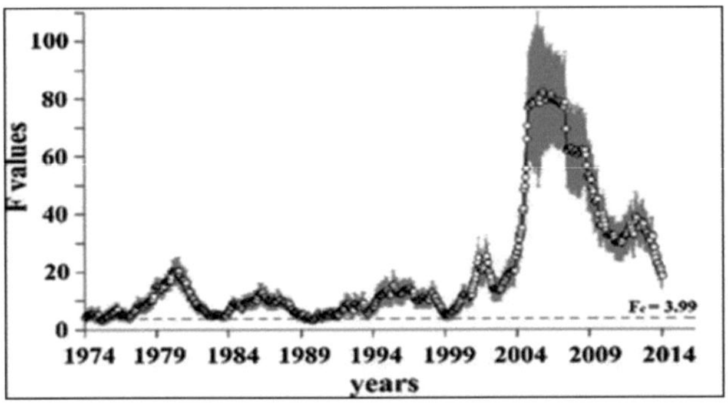

Figure 13.
F-values sequences from the original and the set of randomized seismic catalogs around Enguri dam: the gray bars show the scatted of F and circles are the mean values. The hatched line corresponds to a significant difference between original and randomized catalogs. F-values were assessed using MDs calculated in 50 data windows shifted by one data set.

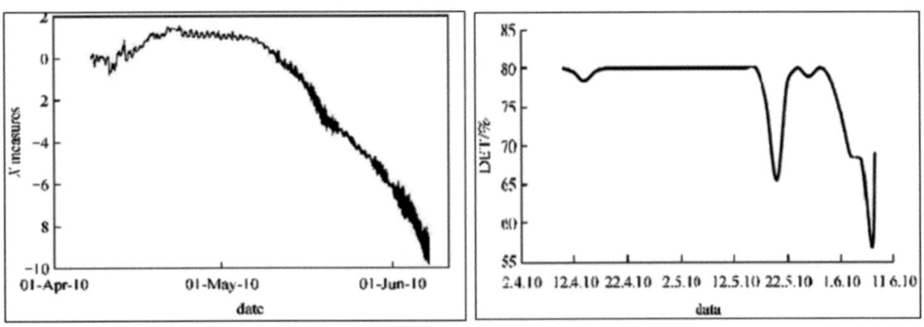

Figure 14.
Results of calculations by DAMTOOL using recurrence quantification analysis (RQA %DET) for the tilt time series from April 2010 to June 2010. Left—original tilt time series; right: RQA%DET (% of determinism) for the same time series. Note high values of DET during a regular regime, which points to a highly correlated regime as well as strong deviations due to geotechnical impact—addition of high-frequency component due to the intensive discharge of water through the dam outlet in two time intervals (12.05–22.05.2010 and 01.06–11.06.2010).

4.5 Module DAMTOOL

For the short-term visual control, processing and nonlinear analysis of monitoring data (say, tilt/strain time series) of engineering constructions the module DAMTOOL was developed. **Figure 14** demonstrates the potential of the nonlinear dynamics approach to the analysis of tilt time series from April 2010 to June 2010. The figure shows the results of calculations by the DAMTOOL package of RQA percent of determinism or RQA %DET. Deviations from the normal behavior in the mid-May and June are evident. Note the high values of %DET during the regular regime in April and first decade of May (which points to the stability of monitoring data) and strong deviations due to geotechnical impacts at the time intervals: 12.05–22.05.2010 and 01.06–11.06.2010, connected with a discharge of water through the dam outlet. This means that construction damage as well as usual geotechnical procedures and geological processes in the dam neighborhood can induce anomalies in the monitoring results. Consequently, it is necessary to keep the minutes of routine geotechnical procedures in order to single out the signatures of hazardous construction damage.

5. Conclusions

The data obtained by the automatic real-time telemetric system of dam diagnostics at the Enguri dam show interesting long- and short-term patterns of tilts/strain dynamics in the dam body. For example, at the highest level (475 m), the maximal displacements of the side sections are larger than critical admissible values (CAVs) by 8–13 mm. This points to the necessity of a new analysis of CAV. Detail observations discover some new facts of dam behavior, including tilt hysteresis during annual loading-unloading cycle, low-frequency dam oscillations, etc., which can be used for dam diagnostics. The possible interpretation of hysteresis phenomena by mesoelasticity (nonlinear elasticity) and thermoelasticity models is considered.

The strain history of the fault zone under the dam during almost half-century documents a persistent extension of fault faces, up to 7500 μm during the observation period as well as dramatic changes in its dynamics with time from 250 to 44 μ /year in the past 20 years.

The nonlinear analysis of tilts/strains time-series data can be used to detect changes in the dam dynamics/stability. The strains of the fault under the dam (FZE) manifest relatively high level of determinism demonstrated by quasi-periodic variations in time due to seasonal load-unload cycle. The FZE is retarded relative to WL maxima by 200 days, which means that the hydraulic diffusivity D of rocks in the fault zone $D \approx 5{\times}10\ m^2/s$.

The authors find three different stages in the nonlinear dynamics of both the strain and seismicity time series: i. before dam erection; ii. during dam construction; and iii. during regular exploitation with characteristic nonlinear dynamics. The start of the ordering process of seismicity (1998–2004) lags behind the fault strain synchronization by water load (1980–1982) by approximately 20 years.

The strains/tilts of the Enguri dam structure as well as that 0f the fault zone under dam reveal complicated dynamics on various time-space scales. Understanding the sources of changes in the FZ dynamics is important for decision makers, who should decide whether the observed data are normal for the routine dynamics of construction or they reveal anomalies, signaling on developing dangerous processes.

Acknowledgements

This work was supported by Shota Rustaveli National Science Foundation (SRNSF), grant FR-19-25231 "New fundamental approaches to assessment of geo-hazards of large hydropower dams/reservoirs: the case of Enguri High Arch Dam Area" and #216732, 2017, as well as by the Council of Europe grant Ref No: GN2019/10; FIMSPO No: 618410 2018-2019.

Author details

Tamaz Chelidze[1,2]*, Teimuraz Matcharashvili[1], Ekaterine Mepharidze[1], Levan Mebonia[3], Mirian Kalabegashvili[3] and Nadezhda Dovgal[1]

1 M. Nodia Institute of Geophysics, Tbilisi State University, Tbilisi, Georgia

2 European Centre "Geodynamical Hazards of High Dams" of the Council of Europe, Tbilisi, Georgia

3 LLC "Engurhesi", Tbilisi, Georgia

*Address all correspondence to: tamaz.chelidze@gmail.com

References

[1] Chelidze T, Matcharashvili T, Abashidze V, Kalabegishvili M, Zhukova N, Meparidze E. Real telemetric monitoring system of large dams and methods for analysis of dam dynamics. Journal of Georgian Geophysical Society. 2011-2012;**15A**:49-66

[2] Perner F, Obernhuber P. Analysis of arch dam deformations. Frontiers of Architecture and Civil Engineering in China. 2010;**4**(1):102-108. DOI: 10.1007/s11709-010-0012-7

[3] Tibaldi A, Alania V, Bonali FL, et al. Active inversion tectonics, simple shear folding and back-thrusting at Rioni Basin. Georgia. Journal of Structural Geology. 2017;**96**:35-53

[4] Chelidze T, Matcharashvili T, Abashidze V, Kalabegishvili M, Zhukova N. Real time monitoring for analysis of dam stability: Potential of nonlinear elasticity and nonlinear dynamics approaches. Frontiers of Structural and Civil Engineering. 2013;7(2):188-205. DOI: 10.1007/s11709-013-0199-5

[5] Hill R. A self-consistent mechanics of composite materials. Journal of the Mechanics and Physics of Solids. 1965;**13**(4):213-222

[6] Chelidze T, Yu K, Matcharashvili T. Seismological criticality concept and percolation model of fracture. Geophysical Journal International. 2006;**164**(1):125-136

[7] Emukhvari N, Bronshtein V. Inguri HPP—System of Allowable and Limiting Parameters of Arch Dam State for Operative Control of its Safety During Exploitation. Moscow-Tbilisi: Ministry of Energy USSR; 1991

[8] Guyer R, Johnson P. Nonlinear Mesoscopic Elasticity. Weinheim, Germany: Wiley-VCH; 2009

[9] Chelidze T, Matcharashvili T, Abashidze V, Tsaguria T, Dovgal N, Zhukova N. Complex dynamics of fault zone deformation under large dam at various time scales. Geomechanics and Geophysics for Geo-Energy and Geo-Resources. 2019;**5**(4):437-455. DOI: 10.1007/s40948-019-00122-3

[10] Sprott JS. Chaos and Time Series Analysis. Oxford: Oxford University Press; 2006

[11] Strogatz S. Nonlinear Dynamics and Chaos. Boulder, Colorado, United States: Westreview Press; 2000

[12] Chen Z, Ivanov P, Hu K, Stanley E. Effect of nonstationarities on detrended fluctuation analysis. Physical Review E. 2002;**65**:041107. DOI: 10.1103/PhysRevE.65.041107

[13] Webber C, Marwan N. Recurrence Quantification Analysis. Heidelberg: Springer; 2015

[14] Zbilut JP, Webber CL. Embeddings and delays as derived from quantification of recurrence plots. Physics Letters A. 1992;**171**:199-203

[15] Hu J, Gao J. Analysis of biomedical signals by the Lempel-Ziv complexity: The effect of finite data size. IEEE Transactions on Biomedical Engineering. 2006;**53**:2606-2609

[16] Li W. Mutual information functions versus correlation functions. Journal of Statistical Physics. 1990;**60**:823-837

[17] Press W, Teukolsky S, Vetterling W, Flannery B. Numerical Recipes: The Art

of Scientific Computing. 3-rd ed. New York: Cambridge University Press; 2007

[18] Vautard R, Yiou P, Ghi M. Singular-spectrum analysis: A toolkit for short, noisy, chaotic signals. Physica D: Nonlinear Phenomena. 1992;**58**: 95-126

[19] McLachlan GJ. Mahalanobis distance. Resonance. 1999;**6**:20-26

[20] Ishimura K, Sakurai S. Effects of window length when smoothing with singular spectrum analysis technique in running data. In: Proceedings of 30th Annual Conference of Biomechanics in Sports. Melbourne: ISBS; 2012

[21] Roy S, Rana D. Machine Learning in Nonlinear Dynamical Systems. Berlin, Germany: Springer; 2021 arXiv: 2008.13496v2

[22] Mata J et al. Validation of machine learning models for structural dam behaviour interpretation and prediction. Water. 2021;**13**:2717. DOI: 10.3390/w13192717

[23] Matcharashvili T, Czechowski Z, Chelidze T, Zhukova N. Changes in the dynamics of seismic process observed in the fixed time windows; case study for southern California 1980-2020. Physics of the Earth and Planetary Interiors. 2021;**319**:45-65. DOI: 10.1016/j.pepi.2021. 106783

[24] Chelidze T, Matcharashvili T. Dynamical patterns in seismology. In: Webber C, Marwan N, editors. Recurrence Quantification Analysis: Theory and Best Practices. Cham: Springer; 2015. pp. 291-335

[25] Chelidze T, Matcharashvili T, Gogiashvili J, et al. Nonlinear analysis of magnitude and interevent time interval sequences for earthquakes of Caucasian region. Nonlinear Processes in Geophysics. 2005;7:9-19

[26] Matcharashvili T, Chelidze T, Abashidze V, et al. Changes in dynamics of seismic processes around Enguri high dam reservoir induced by periodic variation of water level. In: Geoplanet: Earth and Planetary Sciences, Synchronization and Triggereing from Fracture to Earthquake Processes. V. de Rubeis et al. Berlin, Germany: Springer. 2010. pp. 273-286.

[27] Peinke J, Matcharashvili T, Chelidze T. Influence of periodic variations in water level on regional seismic activity around a large reservoir. Physics of the Earth and Planetary Interiors. 2006;**156**(1-2):130-142

[28] Matcharashvili T, Chelidze T, Peinke J. Increase of order in seismic processes around large reservoir induced by water level periodic variation. Nonlinear Dynamics. 2008;**51**:399-407

[29] Telesca L, Matcharasvili T, Chelidze T, Zhukova N. Relationship between seismicity and water level in the Enguri high dam area (Georgia) using the singular spectrum analysis. Natural Hazards and Earth System Sciences. 2012;**12**:2479-2485

[30] Telesca L, Chelidze T. Visibility graph analysis of seismicity around Enguri high arch dam, Caucasus. Bulletin of the Seismological Society of America. 2018;**108**(5B):3141-3147

[31] Biot M. Mechanics of deformation and acoustic propagation in porous media. Journal of Applied Physics. 1962;**33**(4):1482-1498

[32] Shapiro S et al. Characterization of fluid transport properties of reservoirs using induced microseismicity. Geophysics. 2002;**67**(1):212-220

[33] Wang C-Y, Manga M. Earthquakes and Water. Heidelberg: Springer; 2010

[34] Pacheco F. Hydraulic diffusivity and macrodispersivity calculations embedded in a geographic information system. Hydrological Sciences Journal. 2013;**58**:930-944

[35] Bender B, Perkins D. A Seisrisk III: A computer program for seismic Hazard estimation. US Geological Survey Bulletin 1987; 1772. pp. 1-56

[36] Smit P, Arzumanian V, Javakhishvili Z, Arefiev S, Mayer-Rosa D, Balassanian S, et al. The digital accelerograph network in the caucasus. In: The Earthquake Hazard and Seismic Risk Reduction. Netherlands: Kluwer AP; 2000. pp. 85-88

[37] Javakhishvili Z. Seismic hazard and seismicity of Enguri Hydroelectric station area. In: Balavadze B, editor. Geodynamical Studies of Large Dams. Tbilisi, Georgia: Bakur Sulakauri PH; 2002. pp. 88-96

[38] Chelidze T, Matcharashvili T, Abashidze V, Dovgal N, Mepharidze E, Chelidze L. Time series analysis of fault strain accumulation around large dams: The case of Enguri dam, greater Caucasus. In: Bonali FL et al., editors. Building Knowledge for Geohazard Assessment and Management in the Caucasus and Other Orogenic Regions. Berlin/Heidelberg, Germany: Springer; 2020. pp. 185-205

[39] Foulger G, Miles P, Wilson M. Global review of human-induced earthquakes. Earth-Science Reviews. 2017;**178**:438-514. DOI: 10.1016/j.earscirev.2017.07.008

[40] Gupta HK. Reservoir-Induced Earthquakes. Elsevier: Amsterdam, The Netherlands; 1992

[41] Pikovsky A, Rosenblum M, Kurths J. Synchronization: Universal Concept in Nonlinear Science. Cambridge: Cambridge University Press; 2003